FOOD AND BEVERAGE CONSUMPTION AND HEALTH

MOLASSES

FORMS, PRODUCTION AND USES

FOOD AND BEVERAGE CONSUMPTION AND HEALTH

Additional books and e-books in this series can be found on Nova's website under the Series tab.

FOOD AND BEVERAGE CONSUMPTION AND HEALTH

MOLASSES

FORMS, PRODUCTION AND USES

KATELL MADDISON
AND
RANDAL FULLER
EDITORS

nova
science publishers
New York

Library of Congress Cataloging-in-Publication Data

ISBN: 978-1-53614-703-2
Library of Congress Control Number: 2018961872

Published by Nova Science Publishers, Inc. † New York

CONTENTS

PREFACE

Molasses is obtained as a residue of the sugar industry. The major components of molasses are sucrose, glucose and fructose. In Molasses: Forms, Production and Uses, a study is presented wherein polyurethane foams were prepared using the hydroxyl group of mono- and di-saccharides as a reaction site for urethane synthesis. Molasses was dissolved in polyethylene glycol and polyols with various molasses contents were prepared.

The following work contributes to the evaluation of processing technology and the quality characteristics of traditionally made carob molasses through a survey in Tunisia. Carob molasses, known locally as "Rub El Kharroub", is produced mainly by women using an artisanal process carried out with domestic equipment.

Considering that molasses is produced at about 2–5% of the starting raw material, the authors suggest that depending on the raw material condition and applied processing operations, considerable amounts of sucrose can be recovered and an increase in the efficiency of the sugar factory may be achieved.

Following this, the authors review the state of knowledge on the production, chemical composition and uses of sugar cane molasses in animal feeding in Cameroon and briefly examine its other uses. In this country, sugar cane molasses, the main sub-product of sugar industries, is mainly produced by the sugar company in Cameroon.

The authors go on to review the latest advances on the potential of molasses as a source of functional ingredients as well as its application in various food products such as meat, vegetables and fruit. Molasses may have some other valuable functions such as shelf-life improvement, enhancement of leavening activity and buffering capacity.

Lastly, an assessment was made regarding methane production from glycerin digestion and glycerin/molasses co-digestion under thermophilic conditions in a mechanically stirred anaerobic reactor, operated in sequencing batch and fed-batch.

Chapter 1 - Molasses is obtained as a residue of the sugar industry. The major components of molasses are sucrose, glucose and fructose. Polyurethane (PU) foams were prepared using the hydroxyl group of mono- and disaccharides as a reaction site for urethane synthesis. Molasses (ML) was dissolved in polyethylene glycol, and ML polyols with various ML contents were prepared. Using ML polyol, soft-type PU foams, PU composites filled with plant residue having a wide range of functional properties were obtained. Thermal and mechanical properties were con-trolled by changing reaction conditions. Functional properties of PU foams and composites were designed with the purpose of obtaining appropriate products by mixing ML polyols with other types of plant polyols, such as lignin polyol, and plant oil polyol. At the same time, the biodegradability of MLPU foam was examined. Besides PU foams and composites, ML-caprolactone (CL) graft copolymers with various graft ratios were prepared by ring-opening reaction. Molecular length of CL side chain was varied in a wide range and the effect of ML on physical properties was also examined.

Chapter 2 - This work contributes to the evaluation of the process technologies and the quality characteristics of the traditionally made carob molasses through a survey in some Tunisian regions (Monastir governorate). According to the questionnaires, carob molasses, known locally as 'Rub El Kharroub,' has been produced mainly by women using an artisanal process including practically manual operations carried out with domestic equipment. It has been used both in food, as a natural sweetener, and in folkloric medicine as a natural remedy. Four homemade carob molasses provided by different producers were examined for their physicochemical, nutritional, phytochemical and organoleptical properties. The characterization of the samples has revealed that the main physicochemical quality (mainly color and HMF concentration) were related to non- enzymatic browning reactions occurring during juice concentration. Moreover, they serve as a natural source of sugars, minerals and bioactive compounds endowed with biological effects (antioxidant and antibacterial activities). According to the presented results, this study encouraged the consumption of Tunisian carob molasses as a nutritious and healthy food and suggested its use as a functional ingredient in food and pharmaceutical industry.

Chapter 3 - Considering that molasses is produced at about 2 – 5% on the starting raw material (sugar beet or sugar cane) depending on the raw material condition and applied processing operations, considerable amounts of sucrose can be recovered and an increase in the efficiency of the sugar factory achieved. Molasses potential in this field was recognized since sucrose content alongside other sugars accounts for nearly 50%. The presented chapter provides a comprehensive review on sucrose recovery from molasses referred as molasses desugarization process by disclosing common and advanced technologies applied in industry alongside ongoing research in this field. Attention was also directed towards the main results issued from studies on molasses purification by means of standard and emerging filtration aids utilization and application of membrane separation processes as well as combination of these treatments. Particular emphasis was given to recent introduction of bentonite as an adsorbent of molasses non-sugars with the focus on molasses quality enhancement and determination of optimal treatment conditions. A new approach for molasses purification which combines the use of an adsorbent and membrane separation process is also discussed.

Chapter 4 - This chapter reviews the state of knowledge on the production, chemical composition and uses of sugar cane molasses in animal feeding in Cameroon and briefly examine its other uses. In this country, sugar cane molasses, the main sub-product of sugar industries, is mainly produced by the sugar company in Cameroon (SOSUCAM). Formerly used mainly in the fertilization of plantations, it is nowadays a multipurpose by-product. It is used in the production of bioethanol, feed for livestock, pharmaceutical industry and energy production. With a water content of 15 to 25%, molasses contains a large carbohydrate fraction: its non-nitrogenous extract represents about 65% of the fresh product consisting mainly of soluble carbohydrates (58.7%), essentially sucrose (34.6%), glucose (8.5%) and fructose (9.9%). On the other hand, it is deficient in nitrogen (4 to 6%) and has a high mineral content (12 to 16%). It is therefore an energetic food because of the importance of its carbohydrate constituents. It is rich in highly fermentable carbohydrate which stimulates appetite and improves the ingestion of roughage especially in ruminants. In cattle as well as in small ruminants, the synthesis of the results of numerous scientific studies shows that the inclusion of molasses (5 to 10%) in the diet, significantly improves the ingestion and digestibility of the main constituents of the fodder (dry matter, organic matter, proteins, NDF ...). Similarly, the introduction of molasses into the ration has repercussions on zootechnical performance and on the functioning of the gastrointestinal tract of monogastrics. This sub-product, in view of the growing interest in Cameroon's industrial landscape, is nowadays an important resource in terms of animal feed, soil fertilization in plantations and bioethanol production.

Chapter 5 - Molasses is a by-product of industrial processing of sugar from sugarcane or sugar beet. The sugar processing generally consists of serial steps of repeated evaporation, crystallization and centrifugation of cane or beet extraction juices. Molasses remains as dark, viscous syrup following the phases of crystallization and separation of raw sugar. Molasses is a polycomponent system of variable composition due to many factors: biological origin (cane or beet), raw material quality, applied processing methods during juice clarification and sugar crystallization, etc. The main constituents of molasses are fermentable sugars (saccharose, glucose, and fructose) which proportion depends on the nature of molasses (beet or cane). The non-sugar part of molasses is abundant in minerals, especially potassium, sodium, magnesium, calcium and iron, and contains a myriad of versatile other bioactive compounds such as B group vitamins, choline, allantoin, purine, cytosine, guanosine, cytidine, glutamine acid, lactic acid, pectin, phenolics, etc. Beet molasses contains betaine which has been recent-ly recognized as a functional compound with proven beneficial health effect.

A few studies highlighted molasses as a health promoting ingredient due to potent antioxidative activity and proposed its use as a valuable nutritious, yet affordable, commodity. Unlike beet molasses, cane molasses has been used, though not extensively, in bakery and confectionery industry, mainly as a minor ingredient to provide sweetness, unique flavor characteristics and color to a product. Molasses may afford some other valuable functions such as shelf-life improvement, enhancement of leavening activity and buffering capacity. Beet molasses currently has no use in human consumption, mainly due to strong, earthy flavor which is absent in cane molasses. Although odd flavor of beet molasses disables its consumption on its own, several studies have indicated successful use of beet molasses in meat and bakery products. The studies demonstrated the applicable molasses doses and their effects on the various aspects of product quality. It was confirmed that beet molasses can be used in various food products without having adverse effects on product palatability but providing an outstanding improvement in nutrient pattern and antioxidative potential. The aim of the present chapter is to review the latest advances on the potential of molasses (beet or cane) as a source of functional ingredients as well as its application in various food products (bakery, meat, vegetables and fruit products).

Chapter 6 - An assessment was made regarding methane production from glycerin digestion and glycerin/molasses co-digestion, under thermophilic conditions (55°C), in a mechanically stirred (50 rpm) anaerobic reactor, operated in sequencing batch and fed-batch, containing immobilized biomass on polyurethane foam (AnSBBR). The performance of the 5.6 L AnSBBR was assessed at increasing applied volumetric organic loading rate (OLRA) and according to the feeding strategy. During mono-digestion the AnSBBR was batch operated with 2.3 to 6.5 $gCOD.L^{-1}.d^{-1}$ and cycle length of 8 h. During co-digestion (86% glycerin and 14% molasses) the AnSBBR was operated in batch mode with OLRA of 2.4 to 7.7 $gCOD.L^{-1}.d^{-1}$ and cycle length of 8 h, and in fed-batch mode with OLRA of 7.7 $gCOD.L^{-1}.d^{-1}$, cycle length of 8 h and feeding time of 4 h. The best results were obtained during co-digestion with 7.7 $gCOD.L^{-1}.d^{-1}$. In the batch operation, a molar productivity of 84.4 $molCH_4.m^{-3}.d^{-1}$, methane yield of 11.0 $molCH_4.gCOD^{-1}$ with 71.6% of methane in the biogas was obtained. In the fed-batch operation, molar productivity of 83.7 $molCH_4.m^{-3}.d^{-1}$, methane yield 11.2 $molCH_4.gCOD^{-1}$ was obtained with 69.1% of methane in the biogas. There was no significant difference between the results obtained when the feeding strategy was altered, so co-digestion could be performed in both batch and fed-batch with OLRA of 7.7 $gCOD.L^{-1}.d^{-1}$. During mono-digestion hydrogenotrophic methanogenesis predominated up to OLRA of 4.3 $mgCOD.L^{-1}.d^{-1}$. However, when OLRA increased acetoclastic methanogenesis became dominant. During the batch co-digestion hydrogenotrophic methanogenesis was predominant, whereas during the fed-batch co-digestion acetoclastic methanogenesis predominated. The scaling up estimation, considering small and medium-sized biodiesel production industries, resulted in single reactors of 6.64 and 332 m^3, for which parallel operation is suggested with 4 and 8 reactors, with estimated energy generation of 2.94 and 147 kW, respectively.

In: Molasses: Forms, Production and Uses ISBN: 978-1-53614-703-2
Editors: K. Maddison and R. Fuller © 2019 Nova Science Publishers, Inc.

Chapter 1

ADVANCES IN MOLASSES POLYURETHANE FOAMS, COMPOSITES AND MOLASSES-CAPROLACTONE COPOLYMERS

Hyoe Hatakeyama*[*] *and Tatsuko Hatakeyama
Lignocel Research Ltd., Tsukuba, Ibaraki, Japan

ABSTRACT

Molasses is obtained as a residue of the sugar industry. The major components of molasses are sucrose, glucose and fructose. Polyurethane (PU) foams were prepared using the hydroxyl group of mono- and di-saccharides as a reaction site for urethane synthesis. Molasses (ML) was dissolved in polyethylene glycol, and ML polyols with various ML contents were prepared. Using ML polyol, soft-type PU foams, PU composites filled with plant residue having a wide range of functional properties were obtained. Thermal and mechanical properties were controlled by changing reaction conditions. Functional properties of PU foams and composites were designed with the purpose of obtaining appropriate products by mixing ML polyols with other types of plant polyols, such as lignin polyol, and plant oil polyol. At the same time, the

[*] Corresponding Author Email: lignocel@mx3.fctv.ne.jp.

biodegradability of MLPU foam was examined. Besides PU foams and composites, ML-caprolactone (CL) graft copolymers with various graft ratios were prepared by ring-opening reaction. Molecular length of CL side chain was varied in a wide range and the effect of ML on physical properties was also examined.

Keywords: molasses, polyurethane, foams, composites, polyethylene glycol, caprolactone graft-copolymers

Abbreviations

AL	Alcell lignin
CL	ε-caprolactone
C_p	heat capacity, J g^{-1} K^{-1}
DBTDL	dibutyltin dilaurate
DEG	diethylene glycol
ΔH_m	enthalpy of melting, J g^{-1}
ΔC_p	C_p gap at T_g, J g^{-1} K^{-1}
DSC	differential scanning calorimetry
DMA	dynamic mechanical analysis
DMSO	dimethyl sulfoxide
FTIR	Fourier transform infrared spectrometry
HL	hydrolysis lignin
IR	infrared spectrometry
KL	kraft lignin
LS	sodium lignosulfonate
MDI	poly(phenylene methylene) polyisocyanate or 4,4'-diphenylmethane diisocyanate
ML	molasses
MLD	molasses polyol solved in diethylene glycol
MLP	molasses polyol
MLP200	molasses polyol solved in polyethylene glycol 200
MLPU	molasses polyurethane
MLPCL	Molasses-based PCL graft copolymer

M	mass
M_n	number average molecular weight
M_w	weight average molecular weight
PCL	polycaprolactone
PU	polyurethane
PEG	polyethylene glycol
PEG200	polyethylene glycol with molecular mass 200
PPG	polypropylene glycol
T_d	thermal decomposition temperature, °C
TDI	toluene diisocyanate or tolylene diisocyanate
TG	thermogravimetry
T_g	glass transition temperature, °C
T_m	melting temperature °C
WP	wood powder

INTRODUCTION

Molasses is obtained as a residue of the sugar industry. The major components of molasses are glucose, fructose, sucrose and water. The constitution of each component and saccharide content in solution depends on the production site. When saccharide content in molasses is high, the solution is ordinarily converted into alcohol by fermentation. Accordingly, molasses production is markedly affected by trade and the world economy. Production is controlled by a price balance between sugar and industrial alcohol. Several countries control the balance of sugar and alcohol production. However, in countries such as Japan, where sugarcane is not a major agricultural product and industrial alcohol is imported, saccharide content in molasses is reduced as much as possible according to the process, and molasses becomes an industrial residue. In the past, molasses was thrown into the sea, which created a source of water pollution. From the 1980s, efforts have been made to convert industrial plant residues to functional industrial products.

Among a variety of plant residues, molasses has scarcely been paid attention for several reasons. One is mainly due to the above social background based on trade balance. Another is that molasses has not been developed as a resource for industrial products. This may come from a cultural background where there is a demand for alcoholic drinks with high alcohol content. Alcoholic drinks obtained by the fermentation of molasses are limited in terms of world wide scale. Furthermore, in some countries, drinking of alcohol is prohibited for religious reasons, even though sugar-cane is a major product in those countries.

In this chapter, attempts to convert molasses to industrial products will be described based on results obtained by our research group [1, 2]. This chapter consists of two parts. The first part focuses on polyurethane (PU) derived from molasses. Preparation of molasses polyol using polyethylene glycol with various molecular masses receives particular attention. The second part of this manuscript focuses on the development of molasses-based polycaprolactone graft copolymer (MLPCL). Composites using molasses polyurethane or polycaprolactone filled with plant residues are also described.

Soft PU Foams Derived from Molasses Polyol

Preparation of Molasses PU Foams

The major components of molasses are sucrose, glucose and fructose, as shown in Table 1. The amount of each component depends on the sugar production site, and furthermore, different analytical methods sometimes result in inconsistent values. Values in Table 1 show molasses components obtained by a sugar factory in Okinawa (South part of Japan). Molasses was obtained from Tropical Technology Center Co. Okinawa, Japan; the above ML consists of glucose (8.5%), fructose (9.2%), sucrose (32.5), other saccharides (2.3%), ash (9.5%), and water (20.5%). The summation of the percent fraction is not 100%, which is the result of different analytical methods. Molasses was dehydrated before use.

Table 1. Example of chemical components of molasses

Chemical components	%
Sucrose	32.5
Glucose	8.5
Fructose	9.2
Other Saccharides	2.3
Water	20.5
Ash	9.5

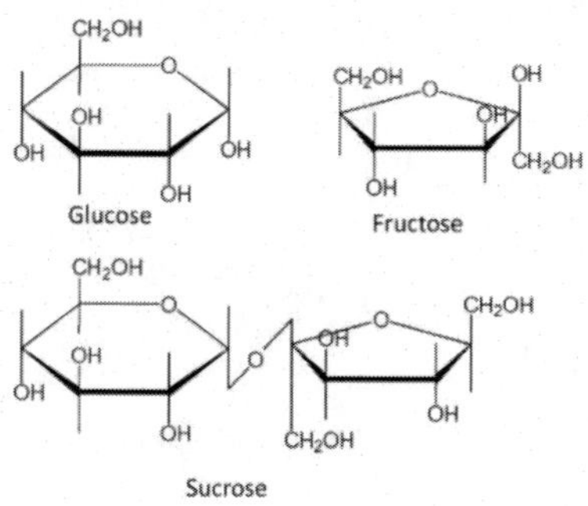

Figure 1. Chemical structure of molasses component.

The chemical structure of mono- and disaccharides is shown in Figure 1. In our previous studies, it was found that mono-and di-saccharides can be solved in polyethylene glycol (PEG) with various molecular masses [3]. Based on this finding, polyols consisting of ML and PEG were prepared. In the first step, dehydrated ML was solved in PEG in various mixing ratios [4, 5], and the preparation conditions were examined. As described in previous reports [1], when ML polyol (MLP) was prepared by mixing 1 portion of dehydrated ML with 2 portions of PEG, the most promising results were obtained. On this account, this mixing ratio was mainly utilized in further reaction processes.

The number of hydroxyl groups in MLP was evaluated according to the methods shown in the Japanese Industrial Standard (JIS) K 1557. The hydroxyl groups in ML reacted with isocyanate via urethane reaction as shown in Figure 2. Through molasses PU preparation, poly(phenylene methylene) polyisocyanate (MDI), toluene diisocyanate (TDI), a mixture of 2,4-TDI (80%) and 2,6-TDI (20%) isomers or MDI/TDI mixtures were used as isocyanate. A synthetic scheme of polyurethane is shown in Figure 2.

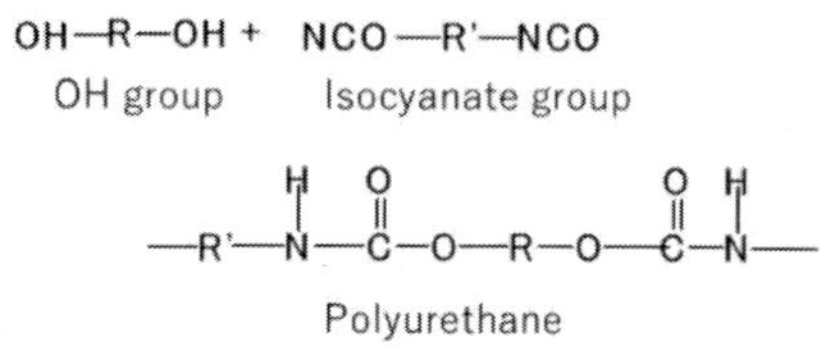

Figure 2. Synthetic scheme of polyurethane. R: saccharide or lignin main structure, R': isocyanate main structure.

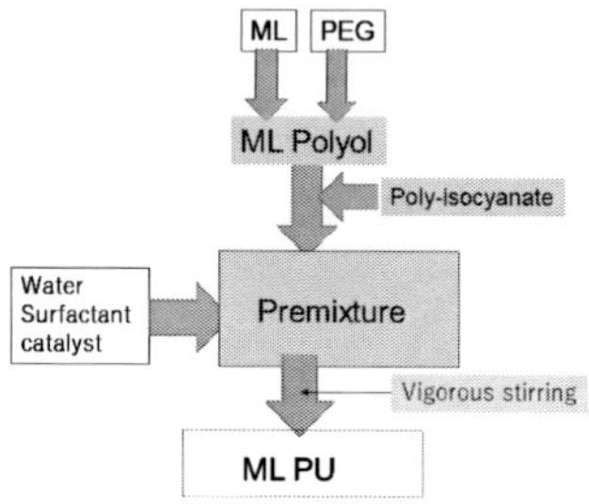

Figure 3. Preparation scheme of MLPU foam. ML; molasses, MLPU; molasses polyurethane.

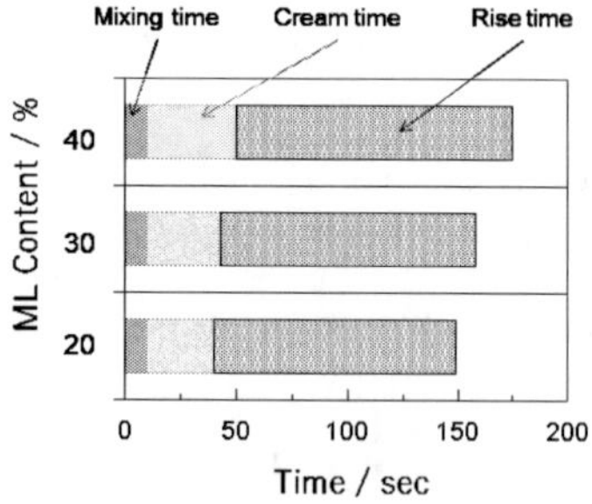

Figure 4. Reaction time of MLPU foam as a function of ML content in polyol. MLP200, MDI, NCO/OH ratio = 1.2.

The reaction was carried out as shown in the schematic flow chart in Figure 3. Mixing conditions of ML polyol, PEG, and kinds of isocyanate were varied taking into consideration the physical properties of obtained PU materials. Polyols, surfactants, and catalysts were homogeneously mixed under stirring, and then isocyanate was added under vigorous stirring. A trace amount of water was used as a foaming agent. NCO/OH ratio was varied from ca. 1.0 to 1.2 whose values were decided in order to fit the physical properties of obtained PU materials. Reaction time is in a range from 2 to 3 min. Several examples of reaction times, mixing time, cream time and rise time [6], are shown in Figure 4. Obtained PU foams were left at room temperature overnight in order to complete the reaction and then used for further investigation.

Through this MLPU preparation, dibutyltin dilaurate (DBDL) was used as a catalyst. Mono- and disaccharides, consisting of ML, have primary and secondary alcoholic hydroxyl groups as shown in Figure 1. Reaction rate of hydroxyl groups in saccharide molecules with isocyanate groups is slow. On this account, DBDL having a high level of activity to the reaction was employed as a catalyst.

When the hydroxyl group of mono- and di-saccharides is used as a reaction site for polyurethane synthesis (Figure 2), it is clear that the above saccharide structure is rigid compared with polyoxyethylene chain. Accordingly, soft type PU foams are formed when MLP is used since MLP has a flexible polyoxyethylene chain. The characteristic feature of MLP is utilized to control physical properties of the other plant polyols which form rigid foams. Control of the rigidity of PU foams will be described in a later section.

A schematic chemical structure of sucrose PU is shown in Figure 5. A photograph of representative MLPU foam is shown in Figure 6, together with a scanning electron micrograph. Information on pore size and the distribution of soft type MLPU foams is found elsewhere [7].

MLPU foams were prepared by changing reaction conditions. Figure 7 shows the relationship between NCO/OH ratio and apparent density of MLPU foam prepared under the same reaction procedure using ML polyol. Apparent density [ρ = mass (m)/apparent volume (V), g cm^{-3}] was measured for samples of 40 (length) x 40 (width) x 30 (thickness) mm using a digital solar caliper and an electric balance. The average values of three samples, whose size was measured at three different spots, were used for the calculation of apparent volume (*V*). The mass (*m*) of each sample was weighed using a balance. As shown in Figure 7, apparent density shows the minimum value at around NCO/OH ratio ca, 1.1 to 1.2, regardless of different type of isocyanate. Based on this result, a further experiment was carried out at NCO/OH ratio = 1.2.

Figure 8 shows apparent density of MLPU foams prepared from MLP with various ML contents. With increasing ML content in polyol, apparent density decreases, regardless of type of isocyanate.

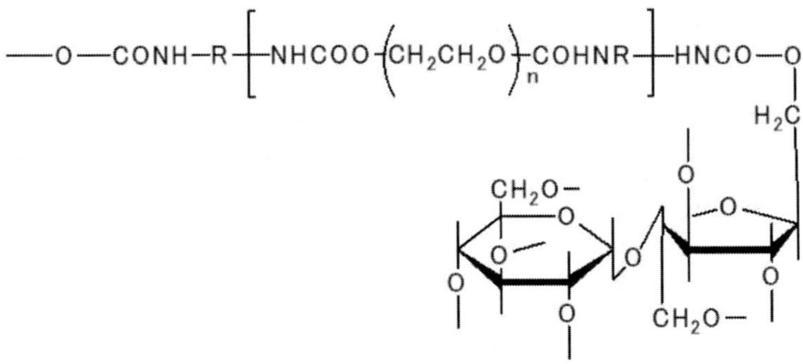

Figure 5. Schematic chemical structure of sucrose PU.

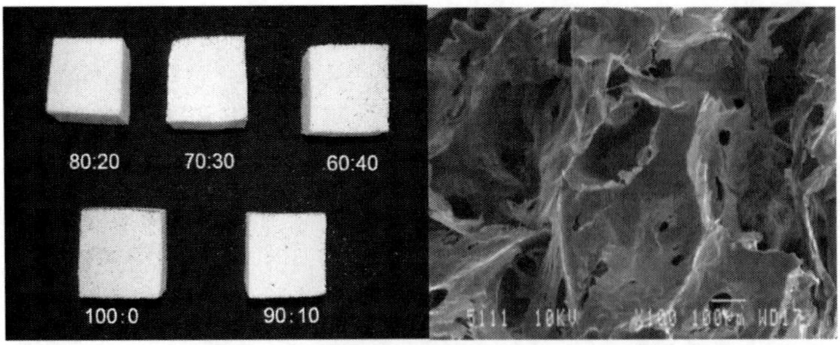

(a) (b)

Figure 6a. Photograph of MLPU. MLP (200), MDI and NCO/OH ratio = 1.05, numerals in the figure show MDI/TDI. Figure 6b. Scanning electron micrograph of surface structure. MDI/TDI ratio = 80:20.

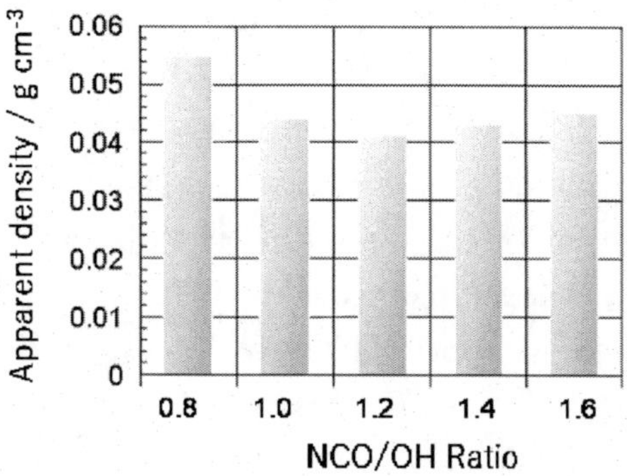

Figure 7. Apparent density of MLPU foams as a function of NCO/OH ratio. MLP (ML = 33%) mixed with MDI.

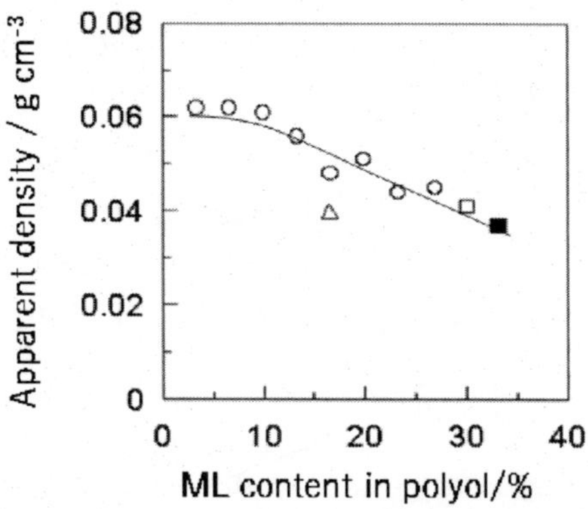

Figure 8. Relationship between apparent density of MLPU foam and ML content in ML polyol. Open circles; open rectangles*; MLP200, NCO/OH ratio = 1.2, MDI *Different source. Open triangle; MLP200/PPG, NCO/OH ratio 1.05, MDI/TDI. Closed rectangle; MLD, NCO/OH ratio = 1.0, MDI.

Properties of Molasses PU Foams

Molecular motion of MLPU is affected by various factors, such as crosslinking density (NCO/OH ratio), molecular mass of PEG, chemical structure of isocyanate, amount of saccharide content and so on. As shown in the schematic chemical structure in Figure 5, the glucose and fructose ring structure acts as a rigid component of PU molecular structure and the linear molecular chain of PEG acts as a soft component. In order to examine the effect of chemical structure on molecular motion of MLPU, glass transition behaviour of MLPU prepared by various conditions was investigated.

It is known that glass transition is a relaxation phenomenon [8]. Molecular chains solidified into glassy state start to move via thermal agitation. Numerous experimental techniques to detect the starting of molecular motion in a long range in polymeric chains are known [9, 10]. Among them, differential scanning calorimetry (DSC) has extensively been utilized due to its reliable performance [9, 11]. By DSC, glass transition (from glassy state to rubbery state) is ordinarily observed as heat capacity (C_p) gap in the baseline, when heating rate is maintained at a constant rate. The rate of molecular enhancement can be estimated via the value of C_p gap (ΔC_p).

Since thermal pre-history affects the glass transition behaviour, MLPU samples were once heated at a temperature 50 °C higher than glass transition temperature (T_g) and then cooled to the glassy state at a constant rate. T_g value of MLPU was determined as schematically shown in Figure 9-(a). Evaluation of T_g by DSC and reliability of data has been reported previously [12]. As shown in Figure 7-b, T_g increases with increasing ML content and T_g value maintains almost a constant value in a ML content from 20 to 33%. Accordingly, in further studies, experiments were carried out using ML polyol with 33% ML.

When PU foams are prepared, the molecular chain of PEG must be chosen taking into consideration the appropriate length. Physical properties of PU foams prepared from different molecular mass of PEG have been extensively investigated. Molecular motion is restricted if intermolecular chain length between rigid structures such as saccharide is short. Accordingly, T_g decreases with increasing molecular mass of PEG [1]. In the case of MLPU, it is appropriate to use short molecular chains as soft components. Generally, when PU foams are derived from plant polyol, they maintain a stable performance, if the number of repeating units (n) of PEG is in a range from 2 to 4 (in a molecular mass from ca. 80 to 160) [1].

However, when MLPU sheet is prepared, PEG with long chains is necessarily required in order to obtain a certain flexibility for sheet processing. Beside MLPU foams, MLPU sheet can also be prepared. An example of PU sheet preparation is as follows [13]. First, ML polyol was prepared by dissolving one portion of ML in 2 portions of PEG with molecular mass = 400 (n = ca. 10). The ML polyol was reacted with MDI at 25°C under stirring until viscous prepolymer was formed. The obtained prepolymer was heat-pressed at 110°C for 5 hours under a pressure of 3 MPa. The PU sheet was cooled at room temperature overnight at the same pressure. The NCO/OH ratio was kept at 1.2.

Glass transition behaviour of MLPU sheets was examined in the same experimental condition as MLPU foams. Figure 10 shows the effect of ML content in polyol on glass transition temperature (T_g) of MLPU sheet. As shown in Figure 10, T_g of MLPU sheet derived from PEG with molecular mass = ca.400 and MDI system without ML is observed at -15°C. When T_g value of MLPU sheet is compared with that of MLPU foam derived from PEG with molecular mass = ca. 200, difference of T_g value is ca. 90°C (Figure10). This indicates that molecular flexibility is strongly affected by molecular chain length of PEG. When ML content in polyol increases, T_g markedly increases. ML acts effectively as a rigid component in MLPU sheet.

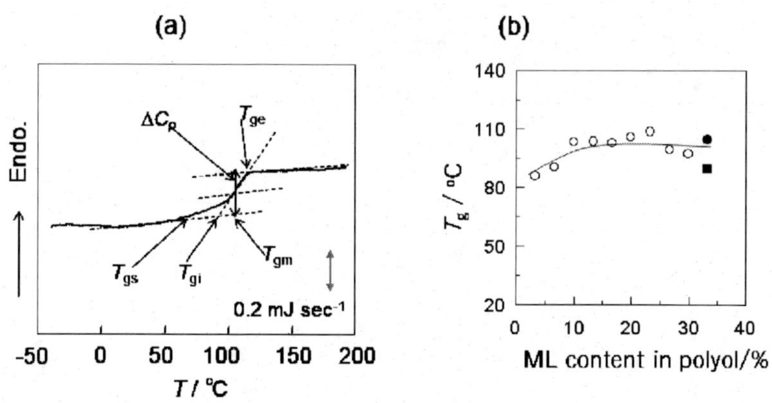

Figure 9. Effect of ML content in polyol on glass transition temperature (T_g). Figure 9a. Schematic DSC heating curve for the evaluation of T_g and ΔC_p (a). Figure 9b. Relationship between T_g and ML content in ML polyol. Open circles; PEG200, MDI, Closed circle; MLD, isocyanate; MDI, NCO/OH ratio = 1.2. Closed rectangle; MLP200 + PPG, MDI, NCO/OH ratio = 1.05. DSC; Heating rate = 10°C min^{-1}, sample mass = ca. 5 mg, N_2 flowing rate = 30 mL min^{-1}.

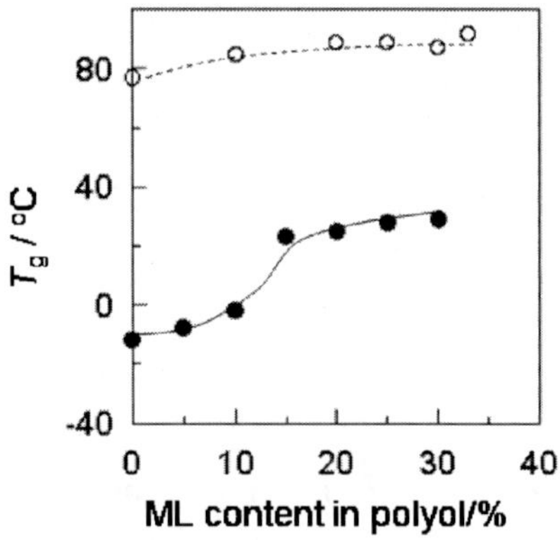

Figure 10. Effect of ML content in polyol on glass transition temperature (T_g) of MLPU sheet. NCO/OH ratio = 1.2, isocyanate; MDI. Closed circles; MLPU sheet (PEG 400, MLP 33%, MDI, NCO/OH ratio = 1.2). Open circles; MLPU foam (PEG 200, MLP 33%, MDI, NCO/OH ratio = 1.2). Seiko DSC220C, heating rate = 10°C min^{-1}, sample mass = ca. 5 mg. N_2 flow rate = 30 mL min^{-1}.

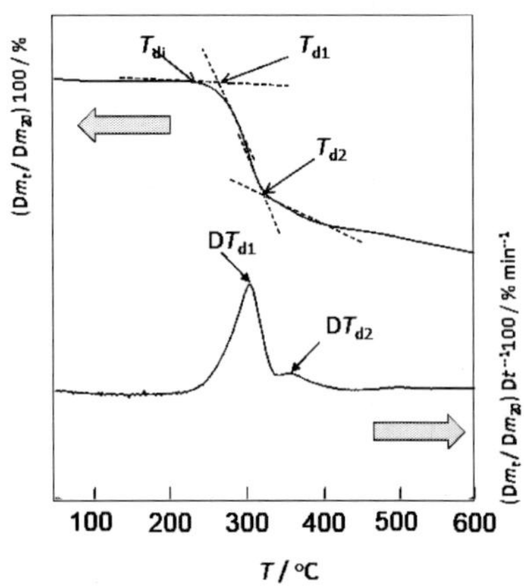

Figure 11. Representative TG curve and DTG curve of MLDPU. Sample; ML content in diethylene glycol = 33%, NCO/OH ratio = 1.2, Isocyanate = MDI. Apparatus; Seiko TG-DTA, heating rate = 20°C min^{-1}, N_2 flow rate = 200 mL min^{-1}, sample mass = ca. 7 mg.

Thermal decomposition mainly depends on chemical structure of PU regardless of macroscopic structure. On this account, thermal stability of MLPU is similar whether MLPU is in foam or sheet form. Thermal decomposition behaviour of MLPU samples was investigated by thermogravimetry

(TG). TG is a technique of thermal analysis (TA), by which mass change is measured as a function of temperature controlled in the constant heating rate [9, 11]. Derivative TG curve is designated as DTG curve.

Figure 11 shows representative TG heating curves of MLPU foam. DTG curves are also shown. Starting temperature of decomposition (T_{di}) and decomposition temperature (T_d) are defined as shown in Figure 11. When several steps of decomposition are observed, T_d was numbered from low to high temperature. The peak temperature observed in DTG curve is numbered, D T_{di} and D T_{d2} as shown in Figure 11. Mass residue (*m*) at a defined temperature (*t*) is also abbreviated as m_t. In this chapter, mass residues at 450°C and 500°C were evaluated.

Representative TG and DTG curves of semi-rigid and soft-type MLPU foams are shown in Figure 12. Curves of A and B show TG and DTG curves of MLDPU prepared from ML polyol with ML content = 33%. Diethylene glycol (DEG) was used as a solvent, poly-MDI was used as an isocyanate, and NCO/OH ratio = 1.2. Curve B shows TG and DTG curves of MLPU foam prepared by the same condition except for NCO/OH. NCO/OH ratio is 1.0. Curve C is MLPU prepared from mixed polyol, which is a mixture of 30% ML polyol, 65% polypropylene glycol (PPG), and 5% diethylene glycol (DEG). Poly-MDI was used as an isocyanate, and NCO/OH ratio = 1.05. ML content in ML polyol was 33% in polyethylene glycol (PEG200). Curve D is MLPU prepared by the same condition as C except for isocyanate. A mixture of MDI and TDI (MDI/TDI = 60/40) was used as an isocyanate.

Mono- and di-saccharides start to decompose at around 200 °C and become caramel. It is appropriate to consider that the low temperature side peak of MLPU is attributable to the decomposition of the saccharide component. Although T_d and the peak temperature of derivative curves of MLPU is higher than that of pure mono- and disaccharides, it is reasonable to consider that the hydroxyl group of saccharides is blocked by urethane linkage. When semi-rigid and soft type PU foams are compared, ML content of soft-type PU foams is small, and accordingly the low temperature side DTG peak in curves C and D is far smaller than that of curves A and D. At the same time, it is known that dissociation of urethane linkage occurs in the temperature range where the first step decomposition is observed. The second step thermal decomposition is attributed to the decomposition of oxyethylene group and that of isocyanurate, which is caused by the ring formation of (NCO–R–NCO) trimers.

Accordingly, the first step thermal decomposition of MLPU foams is attributed to the decomposition of saccharide component of PU. The temperature range of the decomposition can be calculated from peak width of DTG curves. The peak width of the first step decomposition maintained a constant value regardless of MDI content. In contrast, that of the second step decomposition increased with increasing MDI content. The above facts also suggest that the second step decomposition is attributed to isocyanurate.

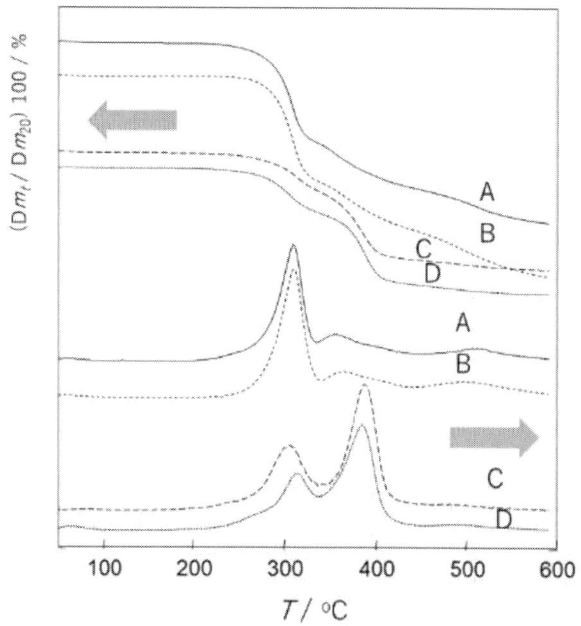

Figure 12. Thermogravimetric curves of MLPU. A; MLDPU, MDI, (ML = 33% in diethylene glycol (DEG), NCO/OH ratio = 1.2, B; MLDPU (ML = 33% in DEG, (NCO/OH ratio = 1.0), C; MLPU (Polyol is a mixture of 30% ML polyol (PEG200), 65% PPG, and 5% DEG, MDI, NCO/OH ratio = 1.05), D; MLPU (Polyol is a mixture of 30% ML polyol (PEG200), 65% PPG, and 5% DEG, MDI/TDI = 60/40, NCO/ OH ratio = 1.05) Apparatus; Seiko TG-DTA, heating rate = 20°C min^{-1}, N_2 flow rate = 200 mL min^{-1}, sample mass = ca. 7 mg.

Table 2. Compression strength (ρ_{10}) and compression modulus of MLPU foams

Sample	Solvent	NCO/OH Ratio	Compression Strength/MPa	Compression Modulus/GPa
MLDPU	DEG	1.0	0.15	3.0
		1.1	0.27	6.8
		1.2	0.31	9.1

ML content in polyol = 33%, DEG; diethylene glycol, isocyanate = MDI, MLDPU; MLPU prepared from MLD Compression strength is obtained at 10% strain [15].

Table 3. Compression strength (ρ_{25}) and compression modulus of soft-type MLPU foams

Sample	Solvent	NCO/OH Ratio	MDT/TDI Ratio	Compression Strength/kPa	Compression Modulus/kPa
MLPU (soft type)	30% MLP, 65% PPG, 5% DEG	1.05	60/40	1.2	6
			70/30	1.5	9
			80/20	2.1	12
			90/10	3.0	20
			100/0	4.1	37

ML content in polyol = 33%, isocyanate = MDI (poly(phenyl methylene) polyisocyanate), TDI tolylene diisocyanate. Compression strength is obtained at 25% strain.

Mechanical properties of MLPU samples depend on the sample processing technique, especially when laboratory scale experiments are carried out. For example, compression strength of MLPU foams derived from a simple combination of MLP + MDI, MLD + MDI has varied in a wide range. After numerous trials, compression strength and modulus of MLPU foams attained satisfactorily acceptable values.

Soft type MLPU foams, which showed a preferable performance, were prepared from a slightly complex system, i.e., polyol was mixed by stirring 30% MLP, 65% PPG, and 5% DEG, and MDI/TDI mixed isocyanate was used (Table 3). Compression strength was measured using a Shimadzu Autograph AG-IS at 25°C. Sample size was 40 (length) x 40 (width) x 30 (thickness) mm. Applied stress was varied in order to control the rate in a range from 1.0×10^{-3} to 1.0×10^{-2} m min^{-1}. Compression strength (σ_{25}, Pa) at the 25% strain was calculated according to JIS JISK6400 and JISK7220. Compression modulus (*E*, Pa) was calculated using the initial stage of stress–strain curve based on JISK7220. Three specimens were examined, and the average value calculated.

Biodegradability of Molasses PU

Various testing methods are used to examine the biodegradability of polymeric materials. Among them, the soil burial method is commonly carried out. Detailed conditions are found elsewhere [1]. MLPU samples were buried in soil in two experimental grounds, one was a nursery garden, and the other a hillside field. 33% MLPU foam test pieces were buried for 12 months in the nursery garden or hillside field. Figure 13a shows a photograph of the nursery garden where the test pieces were buried. Figure 13b shows a scanning electron micrograph (SEM) of the degraded surface of the MLPU foam.

Mass decrease of MLPU in soil is shown in Figure 14a. When MLP is used, mass decrease is accelerated. After 12 months interment, ca, 30% of MLPU mass decreases. ML content in PU also affects mass decrease in soil. Figure 14b shows the comparison of mass decrease among places where soil burial test was carried out. It is clearly seen that the effect of the burial place is an important factor to accelerate degradation. A detailed description concerning testing conditions is found elsewhere [1].

As already shown, MLPU consists of three parts, i.e., saccharides, aromatic compounds from MDI and oxyethylene chains from PEG with various molecular chains. Soil bacteria decompose saccharide into carbon dioxide and water. Oxyethylene chains are decomposed via several kinds of bacteria, such as *Flazvorbacterium sp. Pseudomonas sp. Acinetobacter sp. Aeromonas and Alcaligenes glycororanas.* Aromatic compounds also decompose via several kinds of soil bacteria [16, 17]. A decrease in glass transition temperature of MLPU foams buried in soil was observed by DSC [18].

Figure 13a. Samples buried in the nursery garden. Figure 13b. SEM of 33% MLPU surface buried in the nursery garden for 12 months. scale bar = 10 μm.

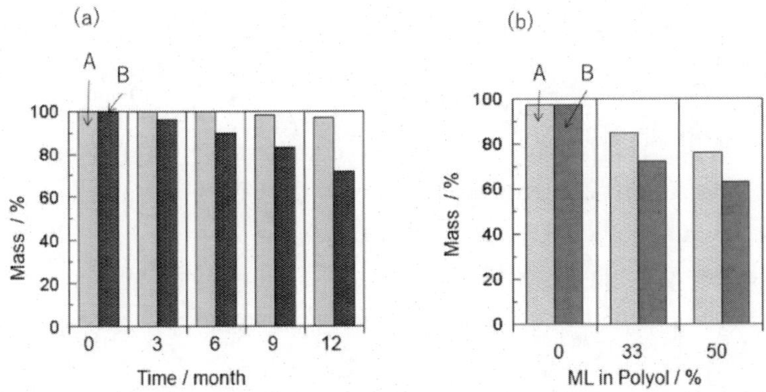

Figure 14a. Mass loss of MLPU foam buried in a nursery garden as a function of time. A; PU without ML, G; PU with 33% ML in polyol. Figure 14b. Mass loss of MLPU foam buried in a nursery garden or hillside field after one year as a function of ML content in polyol. A; PU in nursery garden; PU in hillside field.

PU Foams Derived from Molasses and Lignin Mixed Polyol

Industrial lignin obtained as a by-product of the pulp and paper industry consists of a large amount of green resources which have not been fully utilized, even though numerous attempts have been made to do so [10, 19, 20- 26]. In the course of studies on lignin utilization, various types of industrial products have been developed by a large number of researchers. Among them, PU is the most promising product having a wide range of practical applications. In our study, it is found that lignin is solved in PEG with various molecular masses and PU is successfully prepared using lignin polyol [27, 28]. A wide range of industrial lignins, such as kraft lignin (KL), alcel lignin (AL), hydrolysis lignin (HL), lignosulfonate (LS), etc. have been investigated as a starting material for PU preparation [1, 10, 25, 26]. The characteristic of the above lignin depends on each different separation process of cellulose and lignin. Industrial lignin used in our investigation is shown in Table 3 as a reference. Although the detailed characteristics of industrial lignin vary depending on processing, it is accepted that lignin molecules contain three basic chemical structures, as shown in Figure 15.

Lignin is soluble in PEG with various molecular masses [1]. As shown in the basic chemical structure (Figure 15) [25, 29, 30], the hydroxyl group of lignin is used as a reaction site, when PU is prepared from lignin. As shown in Figure 15, the number of hydroxyl groups of lignin is fewer than that of saccharides (Figure 1). Since the number of reaction sites is fewer, the amounts of soft segments in PU derived from lignin is necessarily small compared with that of MLPU. The above fact suggests that lignin PU is rigid when a comparable reaction procedure is employed (Figure 3) [26]. Characteristic properties of lignin used in this study are shown in Table 4.

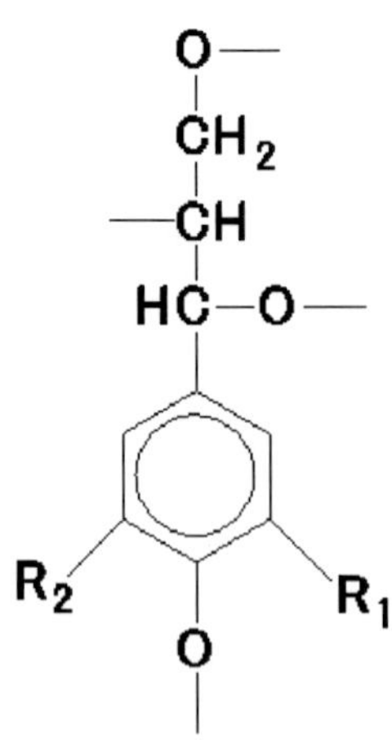

Figure 15. Basic structure of lignin. 1; $R_1 = OCH_3$, $R_2 = OH$: guaiacyl structure. 2; $R_1 = OC\ H_3$, $R_2 = OCH_3$: syringyl structure. 3; $R_1 = OH$, $R_2 = OH$: 4-hydroxyphenylpropane structure.

Table 4. Characteristic properties of lignin used in this study

Lignin	Abbreviations	Characteristics
Kraft lignin	KL	KL was obtained from Westvaco Co. Ltd. Number average molecular mass (M_n) of tetrahydrofuran soluble portion of KL was 1.09 x 10^3 and weight average molecular mass (M_w) was 2.25 x 10^3, accordingly M_w / M_n = 2.03.
Alcell lignin	AL	AL was obtained from Repap Co. Ltd. Number average molecular mass (M_n) of tetrahydrofuran soluble portion of AL was1.0 x 10^3 and weight average molecular mass (M_w) was 2.09 x 10^3, accordingly M_w/M_n = 2.11.
Hydrolysis lignin	HL	HL was obtained from Forestry Technical Academy, St. Petersburg.
Lignosulfonate	LS	LS (commercial name Vanilex HWR) was obtained from Nippon Pulp and Paper Co., Japan. According to the manufacturer's report, molecular mass of LS was 1 x 10^3 - 1 x 10^4.

In order to attain objective physical performance, molasses polyol was mixed with lignin polyol derived from a wide range of industrial lignins, and PU foams were prepared. PU formation using mixed polyols is schematically shown in Figure 16. MLP, in which ML content is 33% and lignin polyol with 33%, was homogeneously mixed and further reaction was carried out. Mixing ratio was varied from 0 to 100%. A detailed description of reaction conditions is found elsewhere [6]. Effect of NCO/OH on physical properties of PU foams derived from ML-Lignin PU foams was investigated [15]. It was found that physical properties show the preferable values when NCO/OH ratio = 1.2. Accordingly, in this text, the results show PU foam prepared at NCO/OH ratio = 1.2. Molecular mass of polyethylene glycol is varied in order to control flexibility of PU foams. Diethylene glycol (DEG) and polyethylene glycol with molecular mass = ca. 200 (PEG200) were mainly used.

A schematic chemical structure of ML-lignin PU is shown in Figure 17.

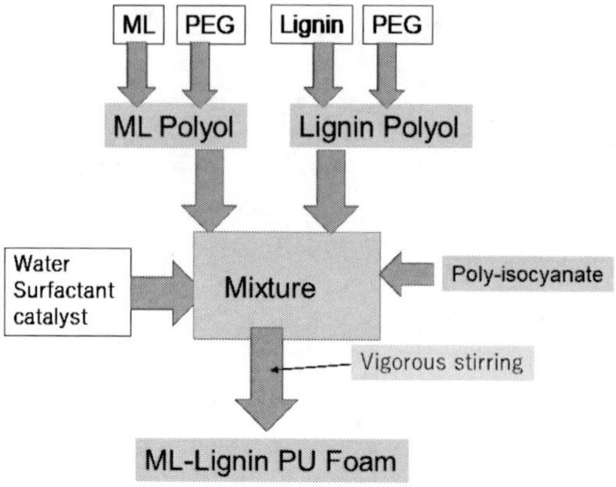

Figure 16. Preparation scheme of ML-lignin PU foam.

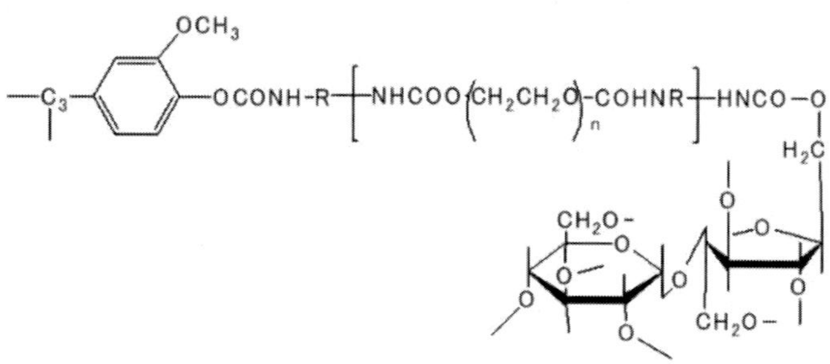

Figure 17. Schematic chemical structure of ML-lignin PU.

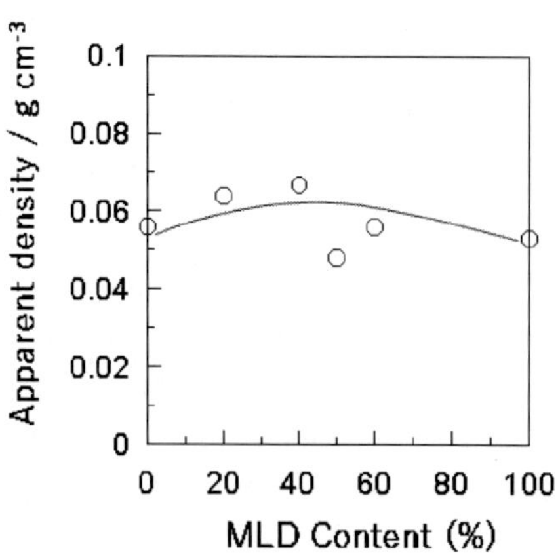

Figure18. Relationship between MLD content in MLD/LSD mixed polyols (%) and apparent density.

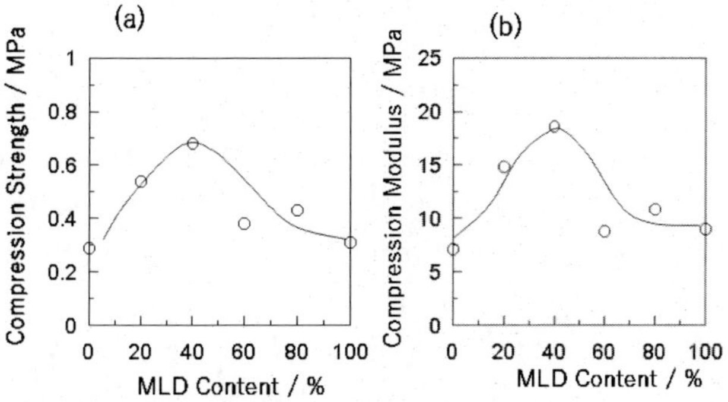

Figure 19a. Relationship between compression strength (σ_{10}) and MLD content in mixed polyols of PU foams derived from molasses-lignin polyol. Figure 19b. Compression modulus (*E*).

The effect of lignin polyol on apparent density of ML polyol is not marked. As shown in Figure 18, apparent density is observed at around 0.06 g cm^{-3}, even if mixing ratio of ML polyol varied from 0 to 100%. In contrast, the effect of mixed polyols on mechanical properties is clearly observed. When polyol derived from LS polyol (ref. Table 3) was mixed with ML polyol, values of compression strength (σ_{10}) and compression modulus (E) markedly changed, as shown in Figures 19a (σ_{10}) and 19b (E). The maximum can be observed in both σ_{10} and E.

The rigidity of molecular chains affects the mechanical response of applied stress and stress recovery, since a wall of PU foam creates a three-dimensional structure. The above facts indicate that mechanical properties can be designed for practical purposes by changing molasses and lignin polyol. By mixing molasses and lignin, rigid or semi-rigid PU foams can be obtained.

PU Foams Derived from Molasses-Castor Oil Mixed Polyol

Castor oil is the standard starting material in the urethane industry and has a long-term historical background. Although castor oil has been converted into chemical products from the petroleum industry, in recent times, plant oil, such as castor oil, soy-bean oil, seed oil, etc. have received renewed attention due to the current circumstances of utilization of biomass for industrial products [2, 31-34]. As already reported, physical properties of PU foam derived from castor oil polyol have been improved by adding MLP [2, 35]. When PU foams are prepared under the same conditions, apparent density decreases from 0.053 to 0.04 g cm^{-3} by adding 30% MLP. Furthermore, glucopyranose and furanose rings of molasses act as a rigid component. Accordingly, by adding 30% MLP to castor oil polyol, glass transition temperature (T_g) of PU foams derived from castor oil increases from 280 to 320 K when PEG with molecular mass 200 was used as a solvent. It was also found that thermal decomposition temperature of castor oil PU shifts to the high temperature side by adding MLP [35].

The major advantage of using MLP is that the physical properties can be controlled in an appropriately planned range. Details of the preparation and physical properties of PU foams derived from molasses-castor oil mix polyol are found elsewhere [2].

Molasses PU Composites Filled with Plant Residue

Recently, various fillers from plant residues have been introduced as fillers for polymeric composites. Plant residues obtained from the pulp and paper industry, bio-diesel production, ethanol fermentation plant, etc. are also used as fillers for polymer composites [for example. 36-39]. In the above studies, matrices of PU composites have been mainly derived from chemicals obtained from petroleum industries. Matrices of composites derived from biomass are important when biodegradability of composites is the main goal [24, 40]. In this section, attempts to make totally "nature friendly composites" having good physical performance are introduced using molasses polyol.

MLPU composites are prepared by adding an appropriate amount of lignin polyol to ML polyol as shown in Figure 20. Wood powder, chitin powder, micro-fibril cellulose having different fibre size, coffee residues (coffee grounds, coffee parchment), oil palm empty fruit bunches etc. are used as a filler. The colour of obtained composites derived from MLPU is in a range from dark yellow to dark brown depending on the kind of filler. Figure 21 shows several examples of MLPU composites filled with wood powder.

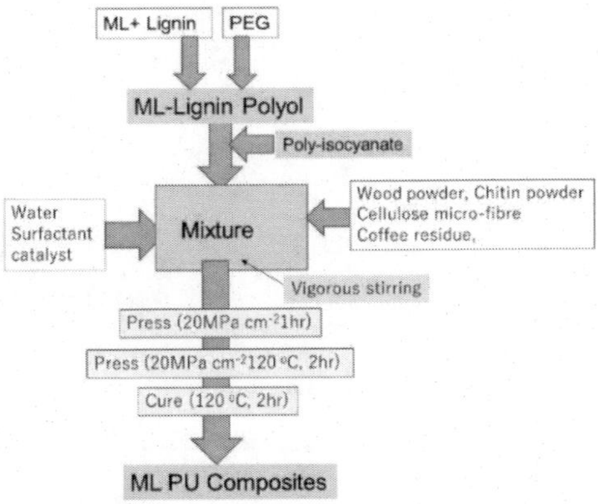

Figure 20. Preparation of MLPU composites filled with natural compounds from plant residues.

Figure 21. Example photographs of ML/LSPU composites filled with wood powder.LSD/MLD = 50/50, scale bar = 30 mm. Numerals show wood powder content (% in polyol).

Figure 22 shows relationships between wood powder (WP) content (%) and apparent density (= ρ, g cm^{-3}) of MLPU-WP composites. Apparent density (ρ) was calculated via ρ = mass/volume using a plate sample with length = 150 mm, thickness =10 mm and width = 30 mm. Apparent density of three different parts of each plate was measured using a caliper and sample mass was weighed. As shown in Figure 22, apparent density of

MLPU composites is constant when wood powder content is the same, regardless of wood powder content in a range from 50 to 70%. At the same time, it is seen that molecular length of ethylene glycol, which acts as a soft segment in the matrix, scarcely affects ρ values.

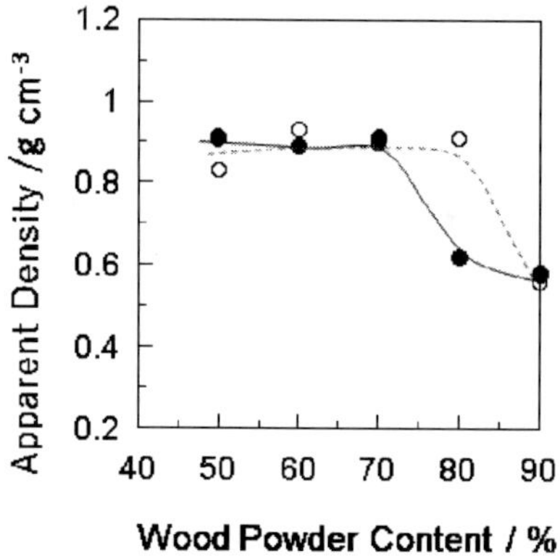

Figure 22. Relationships between wood powder (WP) content (%) and apparent density (g cm^{-3}) of MLPU-WP composites. Open circles; MLD, Closed circles; MLP, ML content of both polyols is 33%, NCO/OH ratio = 1.2. WP used in this experiment was obtained from (Taiyo Chemical Co. Ltd. Wakayama, Japan).

Mechanical properties of MLPU composites were evaluated by bending test at room temperature. As already described in the former section, glass transition temperature of MLPU is observed at around 80°C. Accordingly, MLPU at room temperature, where the bending tests were carried out, is in the glassy state. This indicates that the molecular motion of the matrix is not thermally enhanced at room temperature, suggesting that flowability of composite matrix is restricted. This indicates that bending strength and bending modulus are mainly affected by structural homogeneity of composites. Figure 23 shows relationships between bending strength (Figure 23a), bending modulus (Figure 23b) and wood powder content (%). Bending strength was calculated from stress at strain 10% according to JIS Z2101. Bending modulus was calculated from the gradient of initial linear portion of stress-strain curve. As shown in Figure 25, both bending strength and bending modulus show a maximum at 70 – 80% of wood powder content. It is thought that closely packed condition is attained at around 70%. Furthermore, bending strength and modulus of ML composites prepared from diethylene glycol (DEG) show higher values than those from polyethylene glycol with molecular mass = 200 (PEG). This indicates that selection of molecular chain length is important when mechanical properties are controlled.

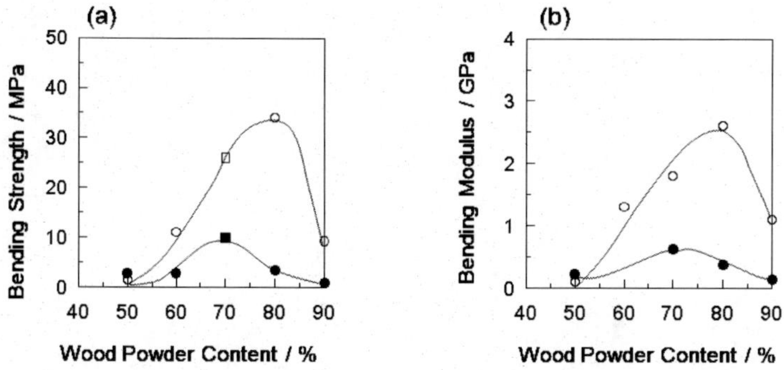

Figure 23a. Relationships between bending strength (MPa) and wood powder content (%). Figure 23b. Relationships between bending modulus (GPa) and wood powder content (%). Open circles. Open rectangle; MLD, Closed circles, closed rectangles; MLP *Circles or rectangles show different experimental sources.

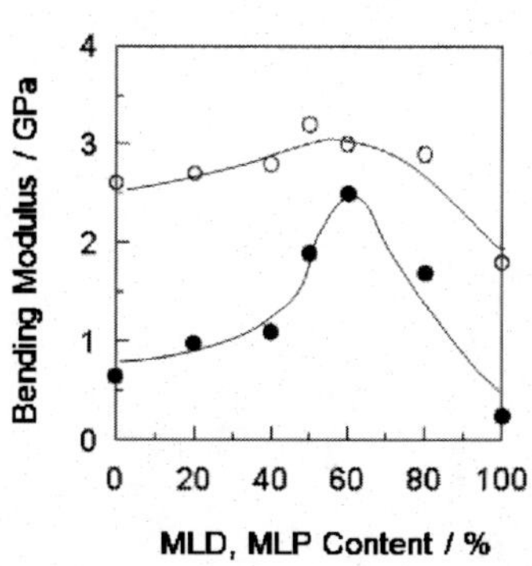

Figure 24. Relationships between MLD, MLP content in MLD/LSD, MLP/LSP polyol and bending modulus of MLPU composites filled with wood powder (70%). Open circles; MLD/LSD, closed circles; MLP/LSP content (%). Characteristics of lignosulphonate (LS) are shown in Table 1.

The effect of mixed polyol on mechanical properties of MLPU composites was also examined. Figure 24 shows relationships between MLD, MLP content in MLD/LSD, MLP/LSP polyol and bending modulus of ML-lignin PU composites filled with wood powder (70%). When the values of binding modulus are compared with MLPU composites with the same amount of filler, bending modulus increases for both series of samples. This implies that the optimum conditions to create better performance in mechanical properties can be attained by appropriate mixing conditions of two types of plant polyol for PU matrix.

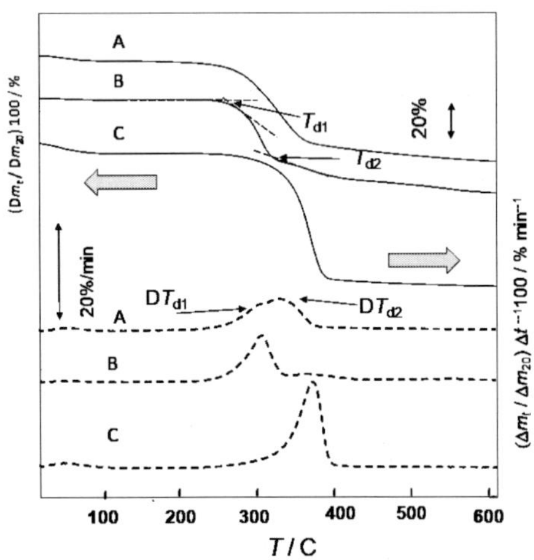

Figure 25. Thermogravimetric (TG) curves and DTG curves of MLPU matrix, MLPU composites filled with wood powder. Solid lines; TG curves, Broken lines; DTG curves. A; MLPU; polyol = MLD, NCO/OH ratio = 1.2, B; MLPU composite filled with wood powder (content = 70%), C; wood powder, T_{d1}; low temperature side thermal decomposition temperature, T_{d2} high temperature side decomposition temperature, DT_{d1}; low temperature side DTG peak of thermal decomposition, DT_{d2}; high temperature side peak. Apparatus; Seiko TG 220, N_2 gas flow rate = 100 mL min^{-1}, heating rate = 20°C min^{-1}, sample mass = ca. 7 mg (each sample was ground, and the obtained powder was packed in a Pt sample pan. temperature range = 25 ~ 600°C.

Figure 25 shows thermogravimetric (TG) curves and DTG curves of MLPU matrix, MLPU composites filled with wood powder (content = 70%) and wood powder. Thermal decomposition temperatures are defined as indicated by the arrows, i.e., T_{d1} is low temperature side, T_{d2} high temperature side decomposition temperature, DT_{d1}; low temperature side DTG peak and DT_{d2}; high temperature side peak, respectively. From this figure, it is clearly seen that the high temperature side decomposition is mainly attributed to filler (wood powder) and the low temperature side decomposition is attributed to matrix MLPU.

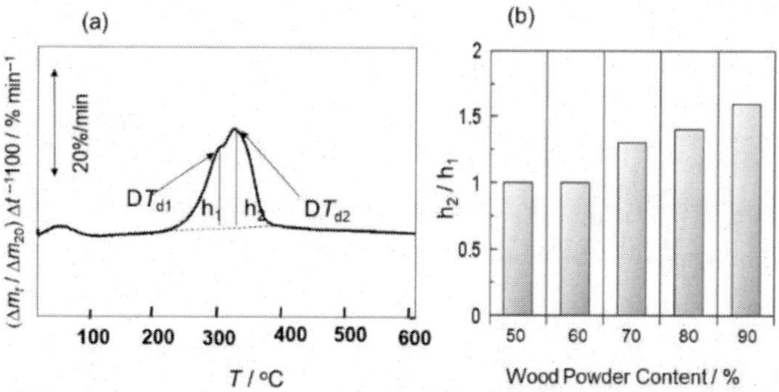

Figure 26a. Definition of peak height of DTG curve of MLPU composite filled with wood powder. Figure 26b. Peak height ratio of DTG peak of MLPU filled with WP (polyol = MLD, NCO/OH ratio = 1.2) as a function of wood powder content.

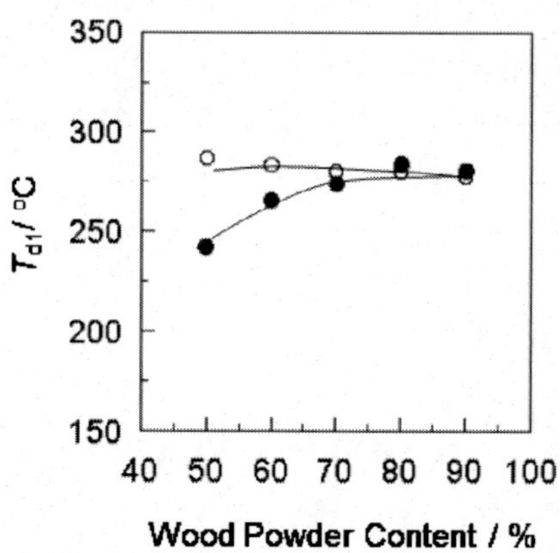

Figure 27. Relationships between T_{d1} and wood powder content. Open circles; MLDPU composite, closed circles; MLPPU. Experimental conditions are shown in the caption in Figure 25.

The effect of filler content in composites on thermal decomposition is clearly seen as a variation of peak height ratio of DTG peak. Peak height ratio in DTG curve (= h_2/h_1) was defined as shown in Figure 26a. In Figure 26b, h_2/h_1 values are shown as a function of filler (wood powder) content. When filler content exceeds 70%, h_2/h_1 values become larger than 1.0. T_{d1} attributed to decomposition of matric (MLPU) is shown in Figure 27.

PU Geocomposites

Geostabilizers are utilized when tunnels are excavated. Urethane injection forepoling is a method used to prevent collapse. The overhead portion and roof of the tunnel are hardened by injection of polyols and isocyanate. Urethane reaction proceeds immediately, and soil and sand are enclosed by PU foam, thus a sudden collapse of the roof of the tunnel is prevented [41]. Excavated sand is piled on the ground or buried in the soil after completion of the construction. Although synthetic stabilizers derived from petroleum are stable in nature and last a long time, if PU from molasses or other natural polyols are used, PU which has acted as a cross linking reagent degrades in soil by the activity of bacterium or microorganisms and decomposes to low molecular mass compounds that take the role of fertilizers in soil. ML polyol, especially ML-lignin polyol is the preferred candidate to convert synthetic geostabilizers to natural ones. Details of geocomposites, preparation and properties using ML polyol can be found elsewhere [1, 42, 43].

MOLASSES-BASED POLYCAPROLACTONE GRAFT COPOLYMERS

Recently "a forgotten polymer polycaprolactone (PCL)" [44] has received attention due to its attractive functionality. PCL, PCL blends and copolymers are thought to be applicable in medical fields, such as tissue engineering, drug-delivery, and scaffolds etc. [45-47] due mainly to their biodegradable nature [48-50]. Synthesis of PCL via polycondensation and ring-opening polymerization has been summarized by Labe [51].

PCL shows high crystallinity and low melting temperature compared with other synthetic polymers. On this account, in the early stage of studies focusing on the improvement of mechanical properties of PCL, chemical modification has been used to increase the processability for a wide range of applications. In order to increase processability, PCL has been blended with amorphous synthetic polymers, such as polystyrene [52], poly(vinyl chloride) [53]. At the same time, PCL was modified by maleate [48] which is an established technique to add matrix polymer viscosity and adhesive properties. The above attempts by blending PCL with synthetic polymer are effective to improve processability and mechanical properties. However, at the same time, the nature friendly capability is reduced by the introduction of synthetic polymers.

Natural polymer-PCL systems have been investigated from the view-point of biocompatible nature. Recently, many researchers have selected chitosan and chitin as a counterpart polymer for PCL blends [54, 55]. It is thought that chitin, which is ordinarily extracted from arthropod, and its derivative chitosan are suitable candidates for formation of new materials for medical applications. Among various kinds of biomaterials, we have extensively investigated industrial materials derived from plant components, including saccharides and lignin [56-59]. Copolymers derived from plant components as a counterpart of PCL have received attention as a represen-tative biodegradable polymer.

In this section, molasses-based PCL graft copolymers (MLPCL) are introduced. Sucrose-based PCL graft copolymer, as a model system of MLPCL, was prepared. The effect of the chemical structure of mono-and oligo-saccharide on the molecular motion of PCL and molecular mass of PCL side chain on crystallization will be described.

Sucrose-Based PCL Graft Copolymer

Sucrose-based PCL graft copolymers were prepared by ring opening reaction as shown in Figure 28. As a model system of molasses, sucrose-based PCL was synthesized in a wide range of caprolactone (CL)/OH ratio as follows. Weighed sucrose and CL were placed in a separable flask with four inlets equipped with Dean-Stark trap. For dehydration, benzene was added and refluxed at 120°C for 1 hr. Benzene was removed by gradual heating, and when the temperature reached 160°C, dibutyltin dilaurate (DBTDL) was added as a catalyst. CL/OH ratio of sucrose was varied from1 to 100 (mol/mol) (1, 2, 3, 4, 5, 10, 15, 20, 25, 30, 45, 55, 60, 70, 100), respectively. The obtained polymer was heat pressed at around 100°C and a sucrose-based PCL sheet was obtained.

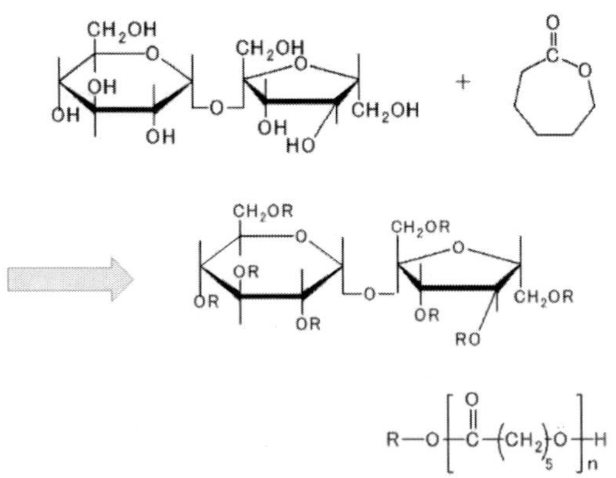

Figure 28. Preparation of sucrose-based PCL graft polymers.

Total conversion, graft ratio and graft efficiency of the reaction system were evaluated as follows:

$$\text{Total conversion } (\%) = (m_{\text{PCL}} / m_{\text{CL}}) \text{ x}100 \quad (1)$$

$$\text{Graft ratio } (\%) = (m_{\text{PCL in graft copolymer}} / m_{\text{sucrose}}) \text{ x}100 \quad (2)$$

$$\text{Graft efficiency (\%)} = (m_{\text{PCL in graft copolymer}} / m_{\text{PCL}}) \times 100 \quad (3)$$

where m_{PCL} is total mass of PCL formed in graft copolymer, m_{CL} is mass of CL monomer, $m_{\text{PCL in graft copolymer}}$ is mass of PCL in graft copolymer and $m_{sucrose}$ is mass of sucrose, respectively. Total conversion of this system is 100%, accordingly, graft ratio linearly increases as shown in Figure 29.

Infrared spectrometry and ^{1}H nuclear magnetic resonance studies on saccharide-PCL copolymer are found elsewhere [60]. Figure 30 shows representative infrared spectrograms of sucrose, PCL and sucrose-based PCL (CL/OH ratio = 30). The absorption band observed at 1720 cm^{-1} is attributed to the ester group. Using the absorption band at 2900 cm^{-1}, which is attributed to the CH group as calibration standard, relative optical density (ROD) was calculated as follows [61]

$$\text{ROD} = (\text{OD at } 1720\ \text{cm}^{-1})/(\text{OD at } 2900\ \text{cm}^{-1}) \quad (4)$$

where OD is base-line optical density.

Relationship between CL/OH ratio and ROD is shown in Figure 31. Values of ROD are constant at around 5.6. The results also verify the result shown in Figure 29.

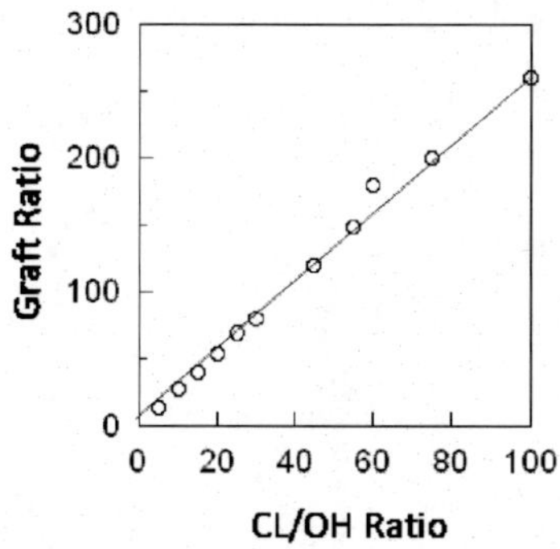

Figure 29. Relationship between CL/OH ratio and graft ratio of sucrose-based PCL.

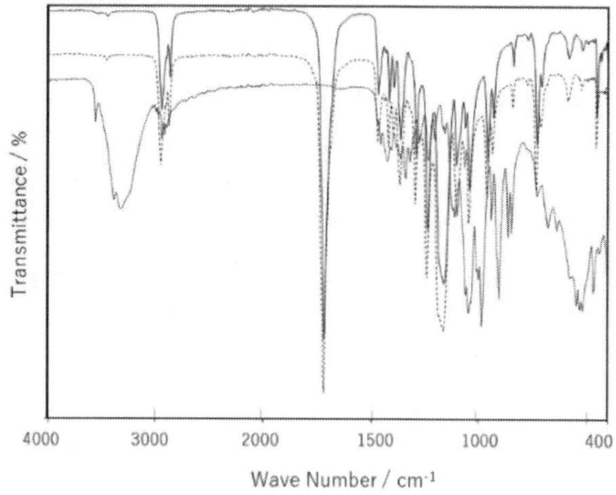

Figure 30. IR spectra of sucose, PCL and sucrose-based PCL. Solid line; sucrose-based PCL (CL/OH ratio = 30), broken line; PCL, Dotted line sucrose. Y axis shifts arbitarilily. Apparatus: Perkin-Elmer IR One.

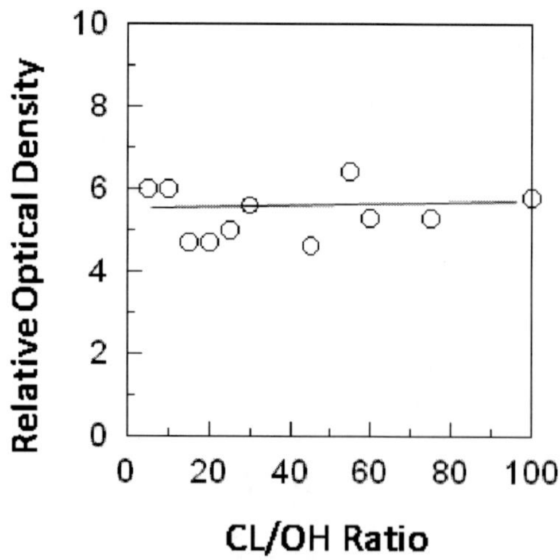

Figure 31. Relationship between CL/OH ratio and ROD = (OD at 1720 cm^{-1})/(OD at 2900 cm^{-1}).

Figure 32 shows representative photographs of sucrose-based PCL graft copolymers. Light yellow sheets can be obtained.

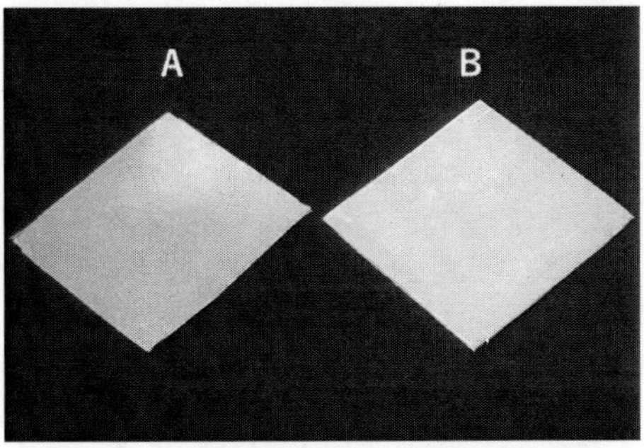

Figure 32. Photographs of sucrose-based PCL graft copolymers. A; CL/OH ratio = 100, B; CL/OH ratio = 75.

Phase transition behaviour of sucrose-based PCL (was investigated by differential scanning calorimetry (DSC). A representative DSC heating curve of sucrose-based PCL is shown in Figure 33a. Phase transition behaviour of crystalline polymer, such as PCL, markedly depends on its thermal history [for example, 59]. The thermal history in a temperature region of crystallization in a cooling process is especially important when the sample is cooled from the molten state. In this study, all sucrose-based PCL samples were melted at 140°C, maintained for 5 min, cooled to -130°C at a cooling rate of 40°C min, and then heated to 140°C again at a heating rate of 10°C min (second heating run). DSC heating curves thus obtained (second heating run) are shown in Figure 33b.

As clearly indicated in Figure 33a, in DSC heating curve of sucrose-based PCL, glass transition was observed as heat capacity gap, and pre-melt crystallization as a shallow exotherm at a temperature lower than melting peak. A large melting endothermic peak was also observed. The melting endotherm shows two peaks. The low temperature side melting peak is designated as T_{ml} and the high temperature side peak as T_{mh}. Melting peaks are difficult to separate into two peaks due to the theoretical background, and accordingly in this text, the peak height shown in Figure 33a was used as a criteria for the magnitude of each peak. From the total peak area, enthalpy of melting (ΔH_m) was calculated. Indium was used as a standard material for calculation. As described in the following section, crystallinity of sucrose-based PCL can be calculated.

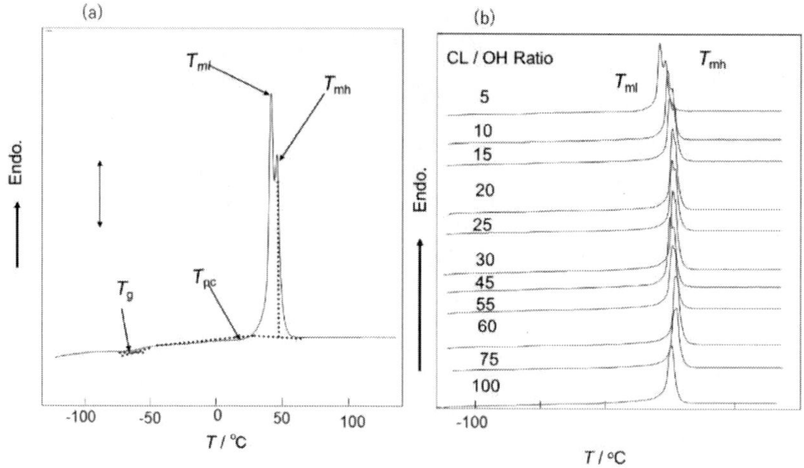

Figure 33a. Representative sucrose-based PCL. T_g; glass transition temperature, T_{pc}; pre-melt crystallization temperature, T_{ml}; low temperature side melting peak temperature, T_{mh}; high temperature side melting peak temperature, H_l; peak height of low temperature side, H_h; peak height of high temperature side. Figure 33b. Stacked DSC heating curves of sucrose-based PCL. Numerals in the figure show CL/OH ratio. Apparatus; Seiko DSC220C, Heating rate = 10 °C min^{-1}, sample mass = ca. 5 mg, N_2 flow rate 30 mL min^{-1}.

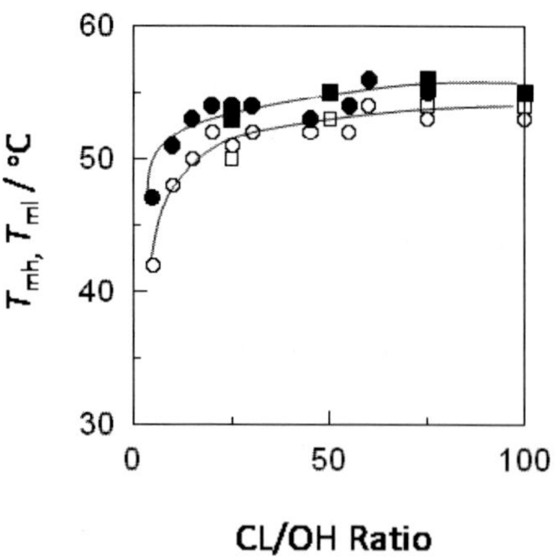

Figure 34. Relationship between CL/OH ratio and melting temperature (T_m) of sucrose-based PCL. Heating rate = 10°C min^{-1}, N_2 flow rate = 30 ML min^{-1}, sample mass = ca. 5 mg. Open circles, open rectangles; T_{ml}, closed circles, closed rectangles; T_{mh} Circles, rectangles; data from different source.

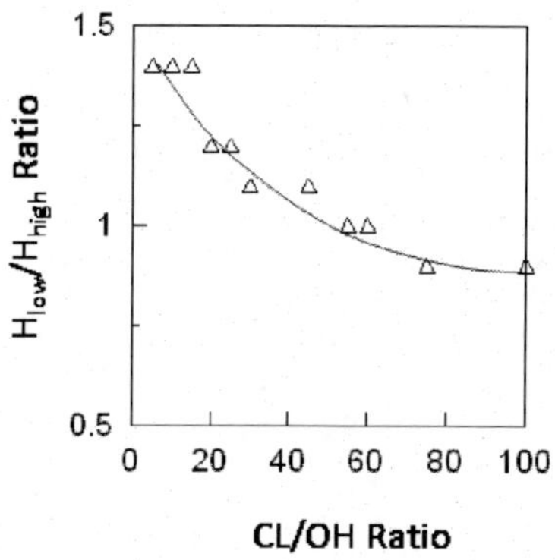

Figure 35. Relationship between CL/OH ratio and melting peak ratio of sucrose-based PCL. H_{low}; peak height of low temperature side, H_{high}; peak height of high temperature side. Heating rate = 10°C min^{-1}, N_2 flow rate = 30 mL min^{-1}, sample mass = ca. 5 mg.

Figure 34 shows the relationship between CL/OH ratio and melting temperature (T_m) of sucrose-based PCL. In this figure, T_m values obtained by other series of experiments are shown together with results obtained from the results shown in Figure 34. Both T_{ml} and T_{mh} increase with increasing CL/OH ratio till ca. 50 and then maintain a constant value. As shown in Figure 33b, the peak height of the low temperature side peak decreases and that of the high temperature side increases with increasing CL/OH ratio. At around CL/OH ratio = 45, it is seen that the high temperature side peak height becomes larger than that of the low temperature side peak.

Peak height ratio (= H_{low}/H_{high}) was estimated using the second heating curve in which the mass of each sample is calibrated. Figure 35 shows the peak ratio of each melting peak as a function of CL/OH ratio. With increasing CL/OH ratio, peak height ratio smoothly decreases. Crystalline structure of PCL is orthorhombic and it is reported that interplanar spacing is affected via the presence of saccharides [55]. Polymorphism is not reported. Accordingly, the low temperature side peak is assumed to be attributed to the crystalline structure with a certain defect caused by sucrose molecules. With increasing PCL portion in graft polymer, PCL molecular chains form a neat crystalline region without the influence of sucrose molecules. On this account, the high temperature side peak is attributable to the melting of crystalline region where PCL molecular chains co-aggregate with no contribution of sucrose molecules.

Figure 36 shows the relationship between CL/OH ratio and melting enthalpy (ΔH_m) of sucrose-based PCL. Melting enthalpy of PCL and PCL graft polymers is markedly affected by thermal history. By slow cooling or annealing at a temperature lower than melting, the high temperature side melting peak increases and at the same time, ΔH_m increases. Accordingly, ΔH_m values shown in this figure show the melting enthalpy of sucrose-based PCL samples melted at 140°C, maintained for 5 min, and cooled to -130°C at a cooling rate of 40°C min. This rapid cooling was intended to detect glass transition clearly. In contrast, it is assumed that reorganization of the molecular chain occurs during heating. Crystallinity calculated by assuming 100% crystalline PCL [49] is shown in Figure 37.

Glass transition of PCL is observed at around a temperature from -60 to -70°C as a gap in the DSC baseline. Glass transition temperature of PCL depends on molecular mass and thermal history which concerns molecular order formation. Molecular motion of amorphous chains is restricted in the presence of the crystalline region. Accordingly, heat capacity difference at T_g (ΔC_p) is thought to be an index of molecular enhancement of molecular chains. The ΔC_p value of PCL and sucrose-based PCL is observed in a range from ca. 0.01 to J $g^{-1}K^{-1}$, which is far smaller than that of completely amorphous synthetic polymer, whose value is observed at around 0.4 J $g^{-1}K^{-1}$. This strongly indicates that molecular movement of PCL molecular chains, thought to be around the sucrose molecules, is restricted by the PCL crystalline region. As shown in Figure 38, T_g value slightly decreases in the initial stage of grafting where PCL molecular chains are arranged randomly. When CL/OH ratio reaches ca. 30, T_g maintains a constant, suggesting that the crystalline region of PCL is well organized. In sucrose-based PCL, molecular motion of sucrose itself is negligible, and no molecular enhancement occurs in the dry state.

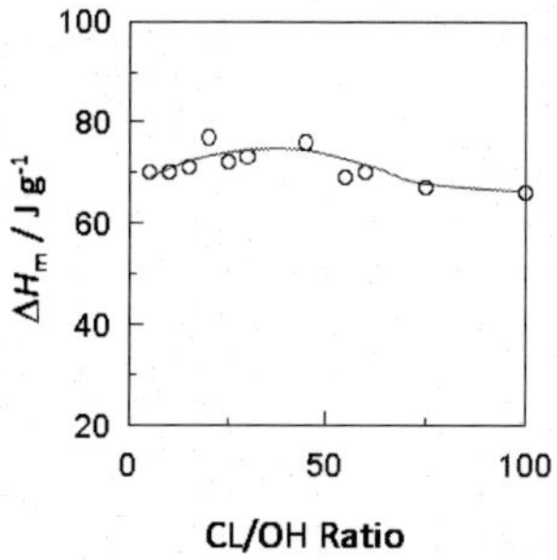

Figure 36. Relationship between CL/OH ratio and melting enthalpy (ΔH_m) of sucrose-based PCL. Heating rate = 10°C min^{-1}, N_2 flow rate = 30 mL min^{-1}, sample mass = ca. 5 mg.

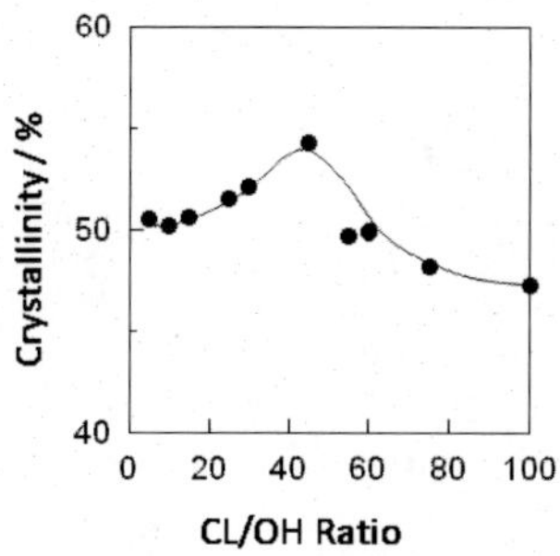

Figure 37. Relationship between CL/OH ratio and crystallinity (%) of sucrose-based PCL.

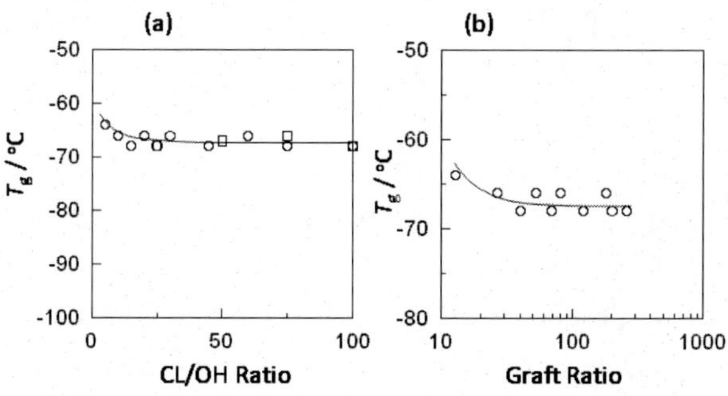

Figure 38a. Relationship between CL/OH ratio and glass transition temperature (T_g) of sucrose-based PCL. Figure 38b. Relationship between graft ratio and glass transition temperature (T_g) of sucrose-based PCL. Heating rate = 10°C min^{-1}, N_2 flow rate = 30 mL min^{-1}, sample mass = ca. 5 mg. Open circles, rectangles circle; data from different sources.

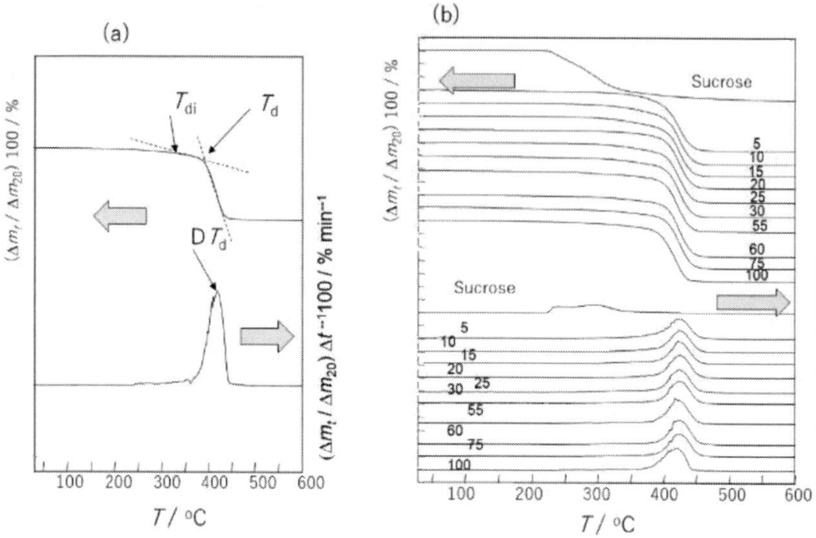

Figure 39a. Representative Thermogravimetric (TG) and DTG curves of sucrose-based PCL and determination of decomposition temperature. T_{di}; starting temperature of decomposition, T_d; decomposition temperature. DT_d; peak temperature of DTG curve. Figure 39b. Stacked TG and DTG curves of sucrose-based PCL with various CL/OH ratios. Numerals in the figure show CL/OH ratio. Apparatus; Seiko TG 220, Heating rate = 20°C min^{-1}, sample mass = ca.7 mg, N_2 gas flow rate =100 mL min^{-1}.

Mono- and oligo-saccharides ordinarily form crystals in dry atmospheric condition. When mono-and oligo-saccharides are quenched from the molten state to room temperature, glassy samples can be obtained, although molecular enhancement is not marked in the dry state. However, in dry state, thermal decomposition starts immediately after melting. Thermal decomposition of sucrose starts at around 200°C as shown in Figure 39b. Accordingly, sucrose is thermally stable at 140°C where PCL and sucrose-based PCL were maintained when DSC measurements were carried out in this experiment.

Analytical data of thermogravimetry (TG) of PCL are obtained from the TG curve as shown in Figure 39a. Change of molecular mass at a temperature (t) was calibrated using mass at 20°C (m_{20}). Starting temperature of thermal decomposition and decomposition temperature are defined in this study as indicated by the arrows. Peak temperature of DTG curve was defined as DT_d. Figure 39b shows TG curves of sucrose-based PCL with various CL/OH ratios. According to Persenaire and his coauthors [62], PCL degrades in a two-stage mechanism. The first process is a statistical rupture of the polyester chains via ester pyrolysis reaction. Produced gases are H_2O, CO_2 and 5-hexenoic acid. The second step is the formation of ε-caprolactone (monomer) via unzipping depolymerization. In TG curves of sucrose-based PCL, two step degradation was not distinctly observed. In Figure 39b, the bottom DTG curve shows sucrose-based PCL with CL/OH ratio 100 in which the contribution of sucrose component is negligible. DTG peak is difficult to separate into two peaks.

Figure 40a shows relationship between CL/OH ratio and decomposition temperature (T_d) of sucrose-based PCL. Magnified curves in a CL/OH ratio from 0 to 25 are also shown in Figure 40b. Td increases in the low CL/OH ratio and then maintains a constant value. Figure 13 indicates that the effect of sucrose on thermal decomposition of sucrose-based PCL is in a limited range, from 0 to ca. 25. Figure 41 shows the relationship between CL/OH ratio and mass residue at 450°C (m_{450}) of sucrose-based PCL. The majority of the sucrose-based PCL decomposes, as shown in Figure 41. Types of evolved gases have been investigated by simultaneous measurement of TG-FTIR [63]. A small increase is observed at around CL/OH ratio= ca. 25, from where the effect of sucrose becomes negligible.

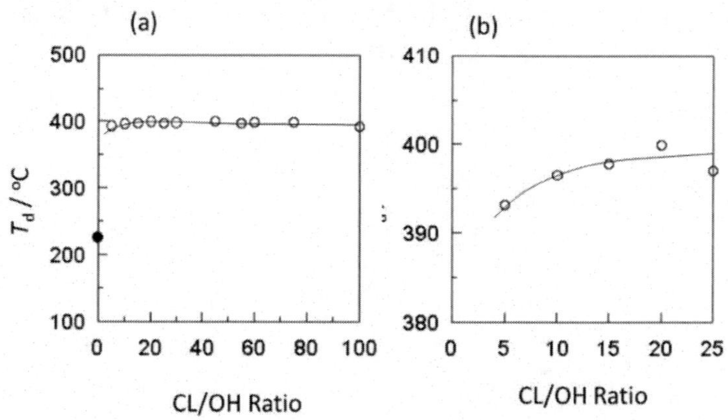

Figure 40a. Relationship between CL/OH ratio and decomposition temperature (T_d) of sucrose-based PCL. Figure 40b. Magnified curve in a CL/OH ratio from 0 to 25. Heating rate = 20°C min^{-1}, N_2 flow rate = 100 mL min^{-1}, sample mass = ca. 7 mg.

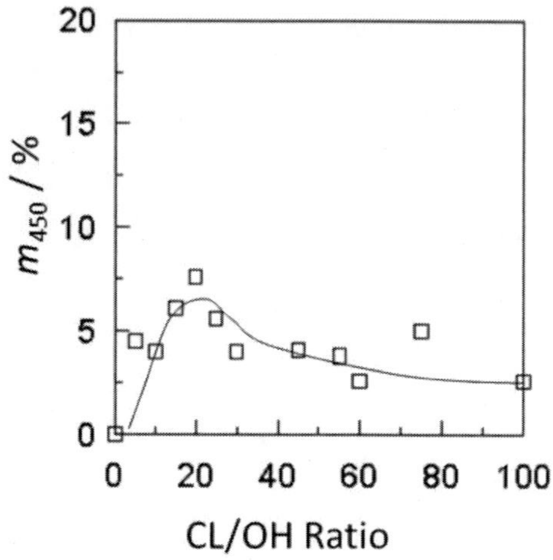

Figure 41. Relationship between CL/OH ratio and mass residue at 450°C (m_{450}) of sucrose-based PCL.

Molasses-Based PCL Graft Copolymer

As described in the former section, molasses obtained as a residue from the sugar industry are a mixture of glucose, fructose and sucrose. Preparation of MLPCL (molasses-based PCL graft copolymer) is similar to sucrose-based PCL, except that dehydration of ML was carried out before reaction. Compared with sucrose-based PCL, reaction sites are complex, leading PCL graft chains to rearrange in a more complicated manner. The DSC melting curves shown in Figure 42 indicate that the low temperature side melting peak maintains a similar size to that of the high temperature side melting peak. In the melting behaviour of sucrose-based PCL, the low temperature side peak is attributed to the melting of crystalline region where saccharide molecules affect the molecular rearrangement of the PCL chain. This suggests that ML components evidently affect the crystallization of PCL.

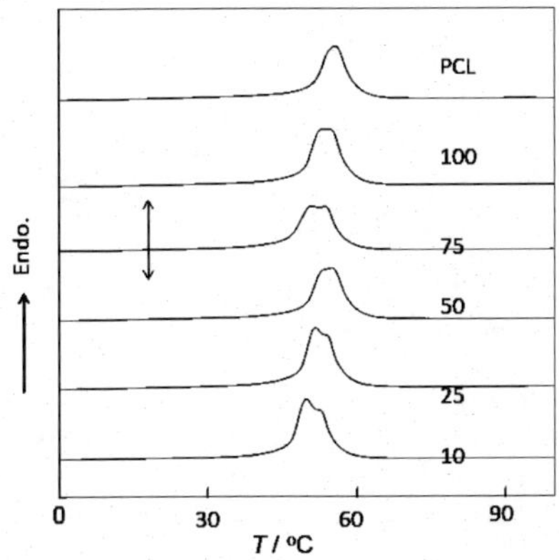

Figure 42. DSC heating curves of MLPCL. Numerals in the figure show CL/OH ratio. PCL is a reference with molecular mass = 5.0 x 10^4 Apparatus: Seiko DSC 220C, heating rate = 10 °C min^{-1}, sample mass = ca. 7 mg, N_2 flow rate = 30 mL min^{-1}.

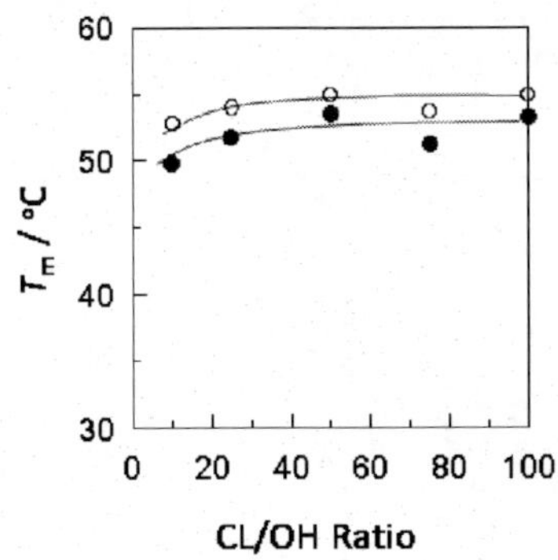

Figure 43 Relationship between CL/OH ratio and melting temperature (T_m) of MLPCL. Heating rate = 10°C min^{-1}, N_2 flow rate = 30 mL min^{-1}, sample mass = ca. 5 mg. Open circles; T_{mh}, closed circles; T_{ml}.

As shown in Figure 43, melting peaks of MLPCL are observed in a temperature range from 50 to 55°C. T_m values of MLPCL are comparable to those of sucrose-based PCL. However, when melting enthalpy (ΔH_m) of MLPCL is compared with that of sucrose-based PCL, ΔH_m values of MLPCL are larger than those of sucrose-based PCL as shown in Figure 44. It is thought that glucose and fructose introduce a bulky space where crystallization is enabled to proceed.

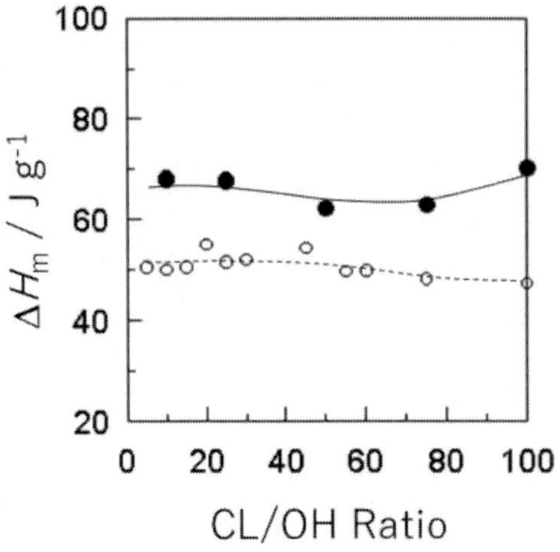

Figure 44. Relationship between CL/OH ratio and melting enthalpy (ΔH_m) of MLPCL. Closed circle; MLPCL, ΔH_m of sucrose-based PCL is shown by open circles for comparison. Heating rate = 10 °C min^{-1}, N_2 flow rate = 30 ML min^{-1}, sample mass = ca. 5 mg.

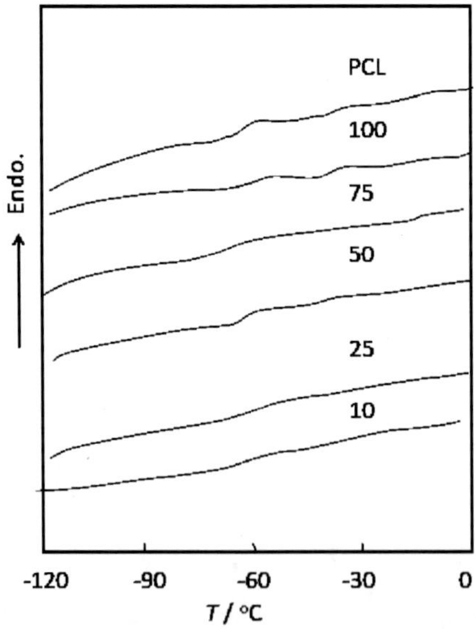

Figure 45. DSC heating curves of MLPCL. Numerals in the figure show CL/OH ratio. PCL is a reference with molecular mass = 5.0 x 10^4 Apparatus; Seiko DSC 220C, heating rate = 10°C min^{-1}, sample mass = ca. 7 mg, N_2 flow rate = 30 mL min^{-1}.

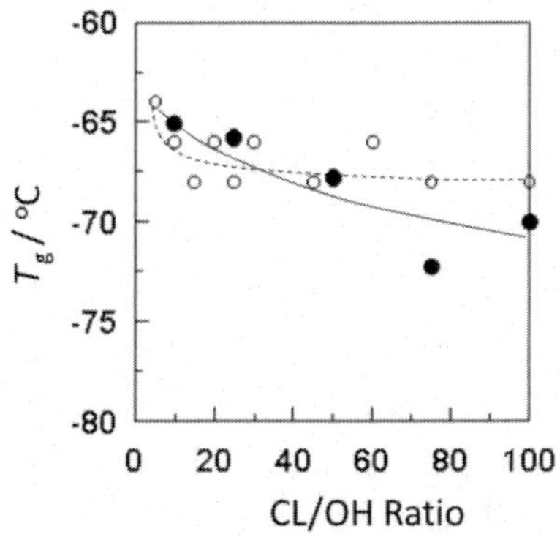

Figure 46. Relationship between CL/OH ratio and glass transition temperature (T_g) of MLPCL Closed circles; MLPCL, open circles; sucrose-based PCL.

Figure 45 shows stacked DSC heating curves of MLPCL in a temperature range of glass transition. Heat capacity gap (ΔC_p) due to glass transition is observable, although the ΔC_p value is not marked compared with values of amorphous polymers [64]. The DSC curve of PCL with molecular mass = 5.0 x 10^4 is also shown. As shown in Figure 46, T_g values of MLPCL and sucrose-based PCL are indistinguishable in a low CL/OH ratio, however, T_g of MLPCL is slightly lower than that of sucrose based-PCL in a high CL/OH ratio. This suggests that saccharides, used as a base of PCL graft chain, affect the molecular motion, even when the length of PCL molecular chain increases.

Figure 47a shows TG and DTG curves of PCL and MLPCL with various CL/OH ratios. T_d and DT_d look identical when curves are compacted and stacked. However, each figure is magnified as shown in Figure 47b, and the difference between MLPCL and PCL is apparent. A shoulder curve can be observed at a temperature lower than DTG peak, indicating that decomposition starts at a lower temperature than that of PCL. Figure 20 shows the relationship between CL/OH ratio and decomposition temperature (T_d) and starting temperature of decomposition (T_{di}) of MLPCL. As shown in Figure 48, thermal decomposition commences at ca. 100°C lower than T_d when CL/OH ratio = 10. With increasing CL/OH ratio, the temperature difference between T_d-T_{di} decreases. Even when CL/OH ratio reaches 100, (T_d-T_{di}) of MLPCL is far larger than that of PCL.

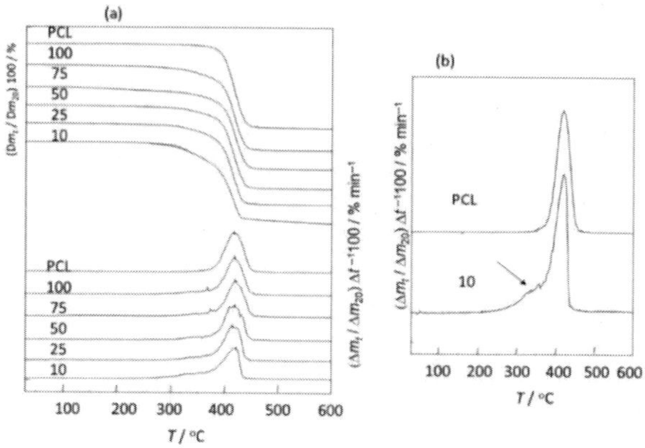

Figure 47a. TG and DTG curves of PCL and MLPCL with various CL/OH ratios. Numerals in the figure show CL/OH ratio. Figure 47b. Magnified DTG curves of PCL and MLPCL with CL/OH ratio = 10. Apparatus; Seiko TG 220, Heating rate = 20°C min^{-1}, sample mass = ca.7 mg, N_2 gas flow rate = 100 mL min^{-1}.

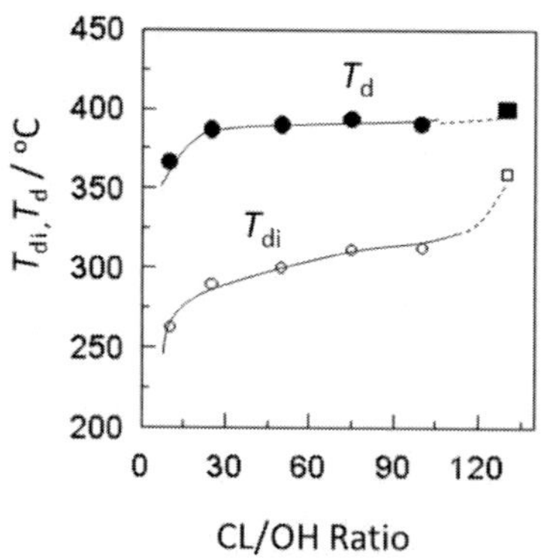

Figure 48. Relationship between CL/OH ratio and decomposition temperature (T_d), starting temperature of decomposition (T_{di}) of MLPCL and PCL. Closed circles; T_d of MLPCL, closed rectangles; PCL, Open circles; T_{di} of MLPCL, Rectangles; PCL, Heating rate = 20°C min^{-1}, N_2 flow rate = 100 ML min^{-1}, sample mass = ca. 7 mg.

Mechanical tests of MLPCL were carried out and stress-strain curves were analyzed according to JISK7113. Tensile strength and tensile modulus of sucrose-based PCL and MLPCL sheets were calculated. Figure 49 shows tensile strength of sucrose-based PCL and MLPCL sheets. Tensile strength is small in the CL/OH region where the saccharide base structure affects the PCL crystallization.

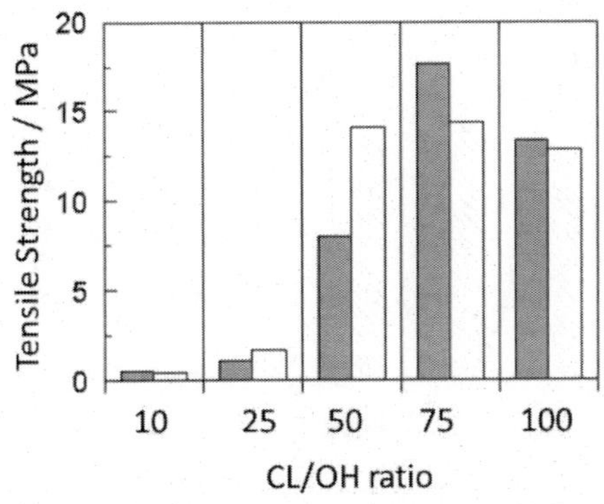

Figure 49. Tensile strength of MLPCL and sucrose-based PCL as a function of CL/OH ratio. Black bar; MLPCL, white bar; sucrose-based PCL.

Sucrose-Based PCL and MLPCL Composites Filled with Chitin and Cellulose

PCL composites reinforced by fillers which are obtained from a wide variety of natural compounds have received particular attention [65]. PCL composites were investigated in order to develop new functional materials intended for medical applications, such as PCL scaffolds reinforced by hydroxyapatite for bone tissue regeneration [66], and chitin-PCL membrane for wound repair [67]. PCL composites were filled with cellulose in order to improve mechanical properties [68]. In this section, sucrose-based PCL and MLPCL composites filled with chitin and cellulose are described. Figure 50 shows the preparation scheme of sucrose-based PCL and MLPCL composites filled with chitin and cellulose.

Filler content was defined as follows:

$$\text{Filler content} = [(\text{mass of filler})/(\text{Mass of PCL graft polymer})] \times 100, \% \quad (5)$$

Filler content was varied from 0 to 40 (0, 5, 10, 15, 20, 25, 30, 35 and 40 %) and CL/OH ratio of matrix PCL graft polymer was maintained at 100. Four series of composites, sucrose-based PCL-cellulose, sucrose-based

PCL-chitin, MLPCL-cellulose and MLPCL-chitin were prepared under the same conditions.

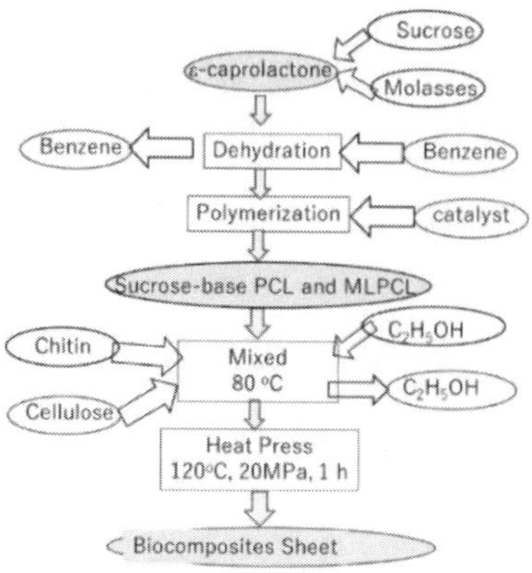

Figure 50. Preparation scheme of sucrose-based PCL and MLPCL composites filled with chitin and cellulose. Sucrose-based PCL and MLPCL were used CL/OH ratio =100.

Figure 51 shows photographs of composites prepared as indicated in the flow chart shown in Figure 50. Both sheets were light yellow. Sucrose-based and MLPCL composites filled with chitin particle (40 %) were also a similar colour. Figure 52 shows a scanning electron micrograph (SEM) of the broken surface of sucrose-based PCL filled with chitin powder (20 %). The size of chitin powder was ca, 80 mesh (180 μm).

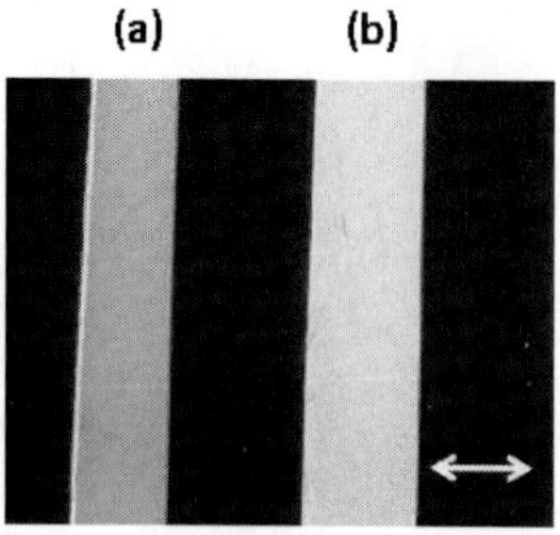

Figure 51. Photograph of sucrose-based (a) and MLPCL (b) composites filled with cellulose particles (40 %). (a); sucrose-based PCL composites (CL/OH ratio = 100), (b); MLPCL composites (CL/OH ratio = 100) The arrow indicates 10 mm.

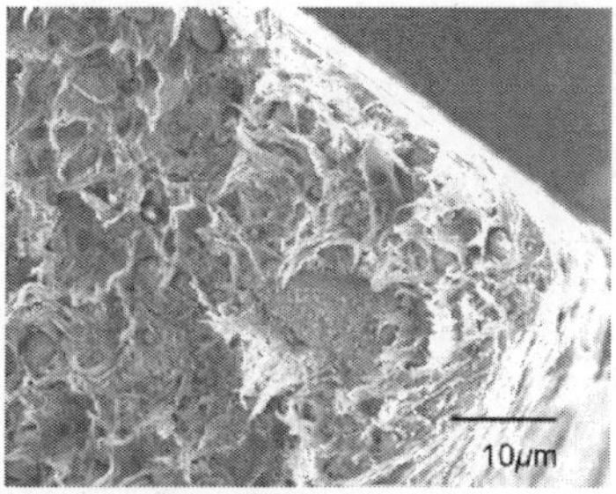

Figure 52. Scanning electron micrograph of broken surface of sucrose-based PCL filled with chitin powder (20%).

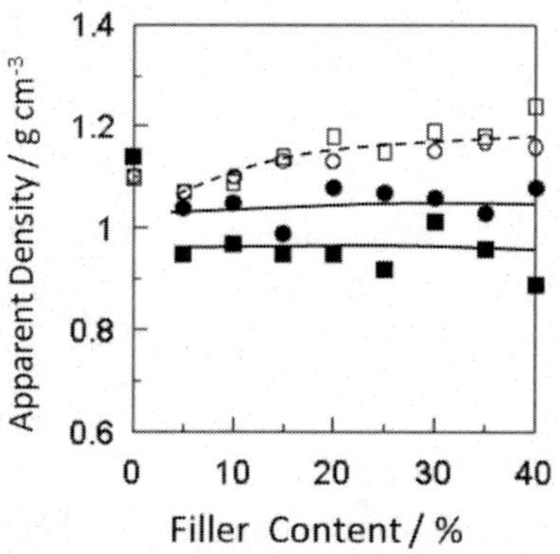

Figure 53. Relationships between filler content and apparent density of sucrose-based and MLPCL composites filled with chitin and cellulose powder. Closed circles; MLPCL composite filled with chitin, closed rectangles; MLPCL composite filled with cellulose, Open circles; sucrose-based PCL composite filled with chitin, Open rectangles; sucrose-based PCL composite filled with cellulose.

Figure 53 shows relationships between filler content and apparent density of composites. When the matrix is sucrose-based PCL, apparent density was scarcely affected by the kind of filler, which was either cellulose or chitin. In contrast, the apparent density of MLPCL filled with chitin is larger than that of MLPCL filled with cellulose. The results indicate that the chemical structure of saccharide affects packing density of fillers, even if the molecular chain length of PCL is the same.

Figure 54 shows DSC heating curves of sucrose-based PCL composites filled with chitin. Chitin content was varied from 10 to 40%. As already described, polysaccharides, such as cellulose and chitin, show no melting in dry state, due to intermolecular hydrogen bonding. Accordingly, melting curves shown in Figure 26 are attributable to the melting of PCL. The low temperature side melting peak is clearly observed. Similar melting curves are obtained for MLPCL filled with chitin. When cellulose was used as a filler, melting curves were found to be similar.

Figure 54 shows DSC heating curves of sucrose-based PCL composites filled with chitin. Chitin content was varied from 10 to 40 %. As already described, polysaccharides, such as cellulose and chitin, show no melting in dry state, due to intermolecular hydrogen bonding. Accordingly, melting curves shown in Figure 54 are attributable to the melting of PCL. The low temperature side melting peak is clearly observed. Similar melting curves are obtained for MLPCL filled with chitin. When cellulose was used as a filler, melting curves were found to be similar.

In the presence of filler, glass transition of composites slightly increases with increasing filler content. As shown in Figure 56, T_g of MLPCL decreases with increasing chain length of PCL. T_g of MLPCL with CL/OH ratio = 100 is ca. -70°C. When chitin content increases in MLPCL matrix, T_g gradually increases as shown in Figure 56. The fact that T_g increases with increasing filler content suggests that amorphous chain movement is slightly restricted via inhomogeneous molecular arrangement.

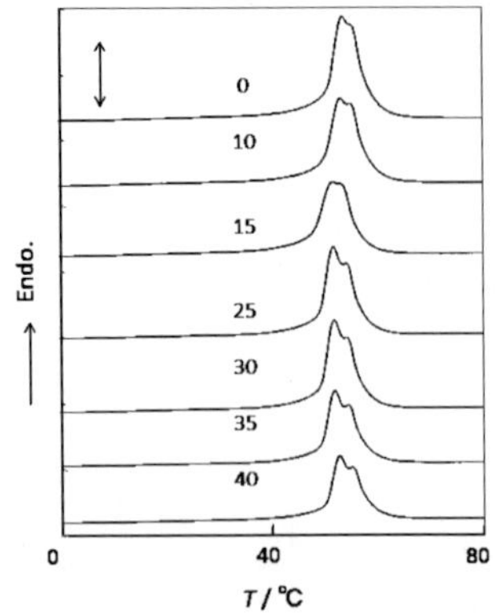

Figure 54. DSC heating curves of sucrose-based PCL composites filled with chitin. Numerals in the figure show filler content. Apparatus; Seiko DSC220C, heating rate = 10°C min^{-1}, sample mass = ca. 7 mg, N_2 flow rate = 30 mL min^{-1}.

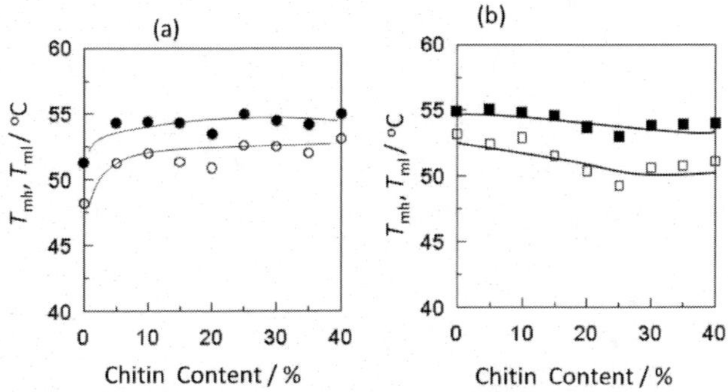

Figure 55a. Relationships between chitin content and melting temperatures (T_{ml} and T_{mh}) of sucrose-based PCL composites (CL/OH ratio = 100). Closed circles; T_{mh}, open circles; T_{ml}. Figure 55b. Relationships between chitin content and melting temperatures (T_{ml} and T_{mh}) of MLPCL composites. Closed rectangles; T_{mh}, open rectangles; T_{ml} Apparatus; Seiko 220C, Heating rate=10°C min^{-1}, sample mass = ca. 5 mg, N_2 flow rate = 30 mL min^{-1}.

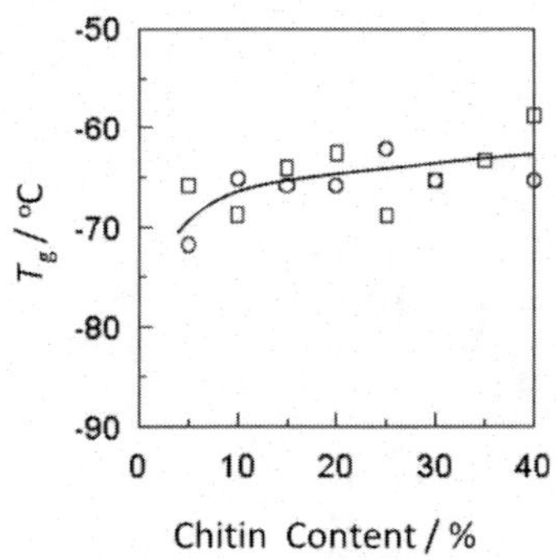

Figure 56. Glass transition temperature (T_g) of sucrose-based PCL and MLPCL filled with chitin. Open circles; sucrose-based PCL composites, open rectangles; MLPCL composites. Heating rate = 10°C min^{-1}.

Thermal decomposition behaviour of MLPCL composites is affected by filler characteristics. TG and DTG curves of MLPCL filled with cellulose powder are shown in Figure 57a and 58a, and those filled with chitin are shown in Figure 57b and 58b. Filler content was varied from 0 to 40%. Two step decomposition was distinctly observed for both series of composites. When DTG curves of MLPCL composites filled with cellulose are compared with those of composites filled with chitin, peak separation of the low and high temperature peak is clearly observed for cellulose filled composites. As shown in Figure 58b, a small shoulder observed in the low temperature side of DTG main peak of MLPCL matrix (the bottom DTG curves in Figure 57b and 58b) is attributable to decomposition of ML. From DTG curves, the starting temperature of decomposition (T_{di}, Ref. Figure 58) of chitin filled composites is observed at a temperature lower than that of cellulose filled composites.

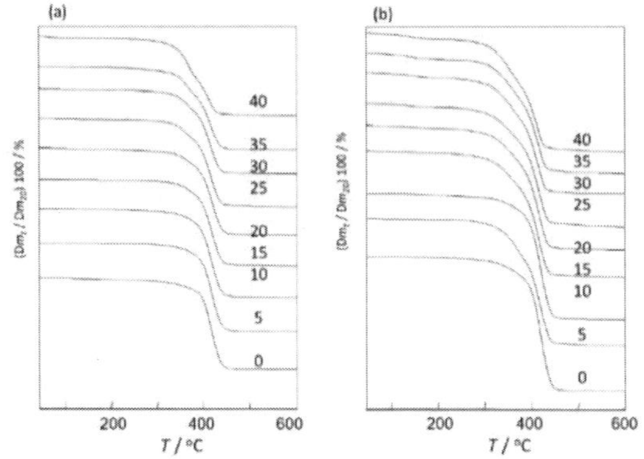

Figure 57a. TG curves of MLPCL composites filled with cellulose. Figure 57b. TG curves of MLPCL composites filled with chitin. Numerals in the figure show filler content. Apparatus; Seiko TG 220, Heating rate = 20°C min^{-1}, sample mass = ca.7 mg, N_2 gas flow rate = 100 mL min^{-1}.

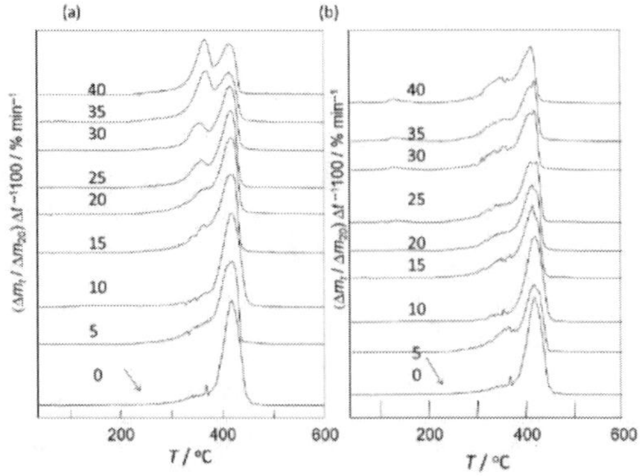

Figure 58a. DTG curves of MLPCL composites filled with cellulose. Figure 58b. DTG curves of MLPCL composites filled with chitin. Numerals in the figure show filler content. Apparatus; Seiko TG 220, Heating rate = 20°C min^{-1}, sample mass = ca.7 mg, N_2 gas flow rate = 100 mL min^{-1}.

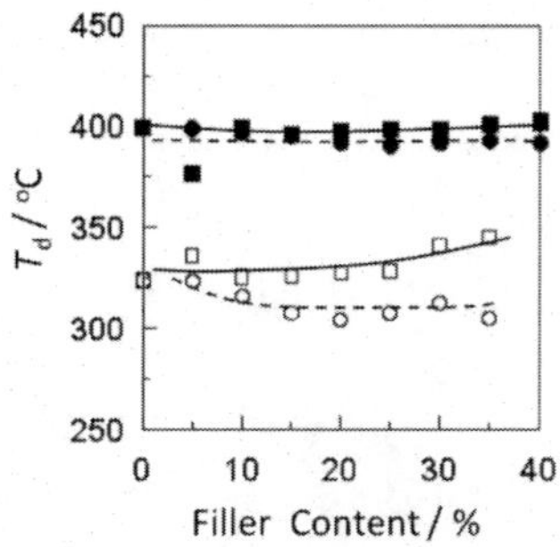

Figure 59. Relationships between filler content and T_d of MLPCL composites filled with chitin or cellulose. Open circles; T_{dL} (chitin), open rectangles; T_{dL} (cellulose) Closed circles; T_{dh} (chitin), closed rectangles; T_{dh} (cellulose).

Figure 59 shows relationships between and T_{dL} and T_{dh} of MLPCL composites filled with chitin or cellulose. Both T_{dL} and T_{dh} of MLPCL composites filled with chitin show a lower temperature than those of cellulose filled composites.

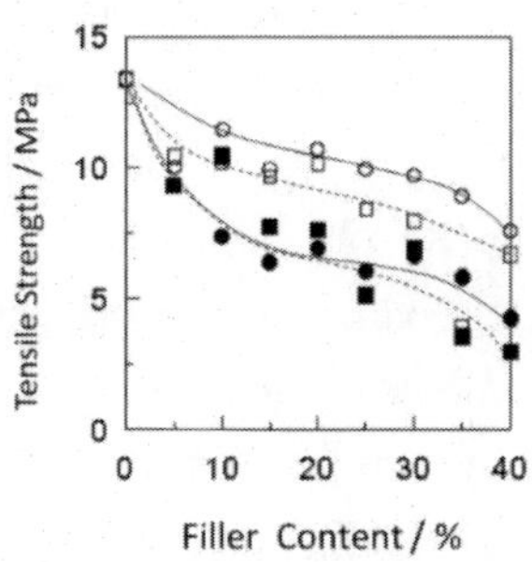

Figure 60. Relationships between filler content and tensile strength of sucrose-based and MLPCL composites filled with chitin and cellulose powder. Closed circles; MLPCL composite filled with chitin, closed rectangles; MLPCL composite filled with cellulose, Open circles; sucrose-based PCL composite filled with chitin, open rectangles; sucrose-based PCL composite filled with cellulose.

Figure 60 shows relationships between filler content and tensile strength of sucrose-based PCL and MLPCL composites filled with chitin and cellulose powder. Mechanical properties of PCL composites decrease with increasing filler content. As shown in this figure, the values of tensile strength of sucrose-based PCL composites are higher than those of MLPCL composites. This suggests that sucrose works as a better compatibilizer in composites.

CONCLUSION

When industrial products are derived from molasses whose major components are glucose, fructose, sucrose and water, it is crucial to take into consideration the fact that the resulting products should have nature friendly characteristics. Future industrial products are expected to merge into biomass circulation in nature i.e., industrial products obtained from plants grown in the presence of water, air, and solar energy, are capable of being biodegraded in soil, and the degraded compounds can be recycled by bacteria and plants.

In this chapter, polyurethanes and polycaprolactones derived from molasses were described. It can be said that industrial products from plant materials have numerous advantages, i.e., not only their biocompatible nature but also the effective utilization of bioresources, which have not been fully utilized. At the same time, when similar products derived from petroleum are compared with those from bioresources, it is not uncommon that products from bioresources lack profitability in the present industrial framework. In order to promote the utilization of bio-compatible products, it is essential to fabricate novel products that meet the needs of the market. From this aspect, in the field of molasses utilization as an industrial product, polyurethanes, especially those derived from molasses-lignin mixed polyol are extremely promising. Polycaprolactone derivatives are an interesting research target, but direct application in a wide range will depend on future investigations.

ACKNOWLEDGMENTS

We wish to express our thanks to Professor Clive S. Langham for his help with the preparation of the manuscript for this article. We also want to extend our thanks to the students at Fukui University of Technology, who were involved in the research and experiments.

References

[1] Hatakeyama, Hyoe, and Hatakeyama, Tatsuko. 2016. *Green polyurethanes: molecular design and application*. New York: Nova Science Publishers, Inc.

[2] Hatakeyama Hyoe, and Hatakeyama, Tatsuko. 2018. "Green Polyurethane Foams and Composites Derived from Castor Oil and Molasses." In *Castor oil and Lesquerella Oils*, edited by Grace Q. Chen, 75-99. New York: Nova Science Publishers, Inc.

[3] Hatakeyama, Hyoe, Hirose, Shigeo., and Nakamura, Kunio. 1997. "Preparation methods of biodegradable polyurethane composites (Patent title: translated from Japanese)." *Japanese Patent* No.2613834 (registered 1997-2-27).

[4] Hatakeyama, Hyoe, Hirose, Shigeo., Kobashigawa, Ken., and Tokashiki, Tadaaki. 1998. "Dehydration of polyols containing molasses with water and polyols containing molasses practically without water (Patent title: translated from Japanese)," *Japanese Paten*t No.2836004, (registered 1998-10-9).

[5] Hatakeyama, Hyoe, Hirose, Shigeo., Kobashigawa, Ken., and Tokashiki, Tadaaki. 1999. "Compositions of polyols for biodegradable soft-type polyurethane foams and the preparation methods of polyurethane foams (Patent title: translated from Japanese)." *Japanese Patent* No. 2883271 (registered 1999-2-5).

[6] Hatakeyama, Hyoe, Hirogaki, Ayako., Matsumura, Hiroshi., and Hatakeyama, Tatsuko. 2013. "Glass transition temperature of polyurethane foams derived from lignin by controlled reaction rate." *J Therm Anal Calorim* 114:1075–1082.

[7] Hatakeyama, Hyoe, Marusawa, Takahiro., and Hatakeyama, Tatsuko. 2011. "Soft type polyurethane derived from molasse." *J Mat Sci* 465: 7475-7481.

[28] Yoshida, Hirohisa, Mörck, Roland., Kringstad, Kunot, P., and Hatakeyama, Hyoe. 1987. "Kraft lignin in polyurethanes I. Mechanical properties of polyurethanes from a kraft lignin-polyethers triol-polymeric MDI system." *J Appl Polym Sci* 34:1187-1198.

[29] Hirose, Shigeo, Yano, Syoichiro, Hatakeyama, Hyoe., and Nakamura, Kunio. 1997. "Preparation methods of novel polyurethanes (Patent title: translated from Japanese)." *Japanese patent* No. 2675997 (registered 1997-7-18).

[30] Lin, Stephen Y., and Dence, Carlton W. ed. 1992. *Methods in Lignin Chemistry*, Berlin: Springer-Verlag.

[31] Lapierre, Catherine. 2010. "Determining Lignin Structure by Chemical Degradation." In *Lignin and Lignans, Advances in Chemistry*, edited by Cyril Heitner, Donald R. Dimmel, and John A. Schmidt, 11-48. Boca Raton: CRC Press.

[32] Montero de Espinosa, Lucas, and Meier, Michael A. R. 2011. "Plant oils: The perfect renewable resource for polymer science?!" *Eur Polym J* 47: 837-852.

[33] Narine, Suresh S., Yue, Jin, and Kong, Xiohua. 2007. "Production of polyols from canola oil and their chemical identification and physical properties." *J Am Oil Chem Soc* 84: 173-179.

[34] Fan, Hongyu, Tekeei, Ali., Suppes, Galen J., and Hsieh, Fu-Hung. 2013. "Rigid polyurethane foams made from high viscosity soy-polyols." *J Appl Polym Sci* 127: 1623-1629.

[35] Guo, Andrew, Javni, Ivan., and Petrovic, Zoran. 2000. "Rigid polyurethane foams based on soybean oil." *J Appl Polym Sci* 77: 467-473.

[36] Hatakeyama, Hyoe, Matsumura, Hirosi., and Hatakeyama, Tatsuko. 2013. "Glass transition and thermal degradation of rigid polyurethanes foams derived from castor oil-molasses polyols." *J Therm Anal Calorim* 111:1545-1552.

[37] Fuad, Ahmad M. Y., Mustafah, J., Mansor, M. S., Ishak, Mohd Z.A., and Omar, Mohd, A. K. 1995. "Thermal properties of polypropylene/ rice husk ash composites." *Polym Inter* 38: 33-43.

[38] Gauthier, R., Joly, C., Coupas, A. C., Gauthier, H., and Escoubes, M. 1998 "Interfaces in polyolefin/cellulosic fiber composites: chemical coupling, morphology, correlation with adhesion and aging in moisture." *Polym Compos* 19: 287-300.

[39] Le Digabel, F., Boquillon, N., Dole, P., Monties, B., and Averous, L. 2004. "Properties of thermoplastic composites based on wheat-straw lignocellulosic fillers." *J Appl Polym Sci* 93: 428–436.

[40] Dányádi, Lívia, Renner, Károly., Móczó, János., and Pukánszky, Béla. 2007. "Wood flour filled polypropylene composites: Interfacial adhesion and micromechanical deformations." *Polym Eng Sci* 47:1246-1255.

[41] Hatakeyama, Hyoe, Ohsuga, Tadahide., and Hatakeyama, Tatsuko. 2014. "Thermogravimetry on wood powder-filled polyurethane composites derived from lignin." *J Therm Anal Calorim* 118:23–30.

[42] Higo, M., Nagasawa, N., Inoue, K., Maki, Furukawa, K., and Nakagawa, K. A study on characteristics of material and form of injection for urethane injection forepoling in tunneling. *North American Tunneling 96*, Ozdemir ed. 1996 Balkema, Rotterdam.

[43] Hatakeyama, Hyoe, Hirose, Shigeo, Wakisaka, Osamu., Mizuno, Noboru., and Ikaga, Syuji. 2004. "Binders for geocomposites (Patent title: translated from Japanese)." *Japanese Patent* No. 3530977 (registered 2004-3-12).

[44] Hatakeyama, Hyoe, Nakayachi, Akinobu, and Hatakeyama, Tatsuko. 2005. "Thermal and mechanical properties of polyurethane-based geocomposites derived from lignin and molasses." *Comp Part A: Applied Sci & Manufac* 36:698-704.

[45] Woodruff, M. A., and Hutmacher, D. W. 2010. "The return of a forgotten polymer-Polycaprolactone in the 21st century." *Prog Polym Sci* 35:1195-1216.

[46] Pitt Colin G. 1990. "Poly-ε-caprolactone and its copolymers." In "*Biodegradable polymers as drug delivery systems*," edited by Chasin Mark, Langer Robert, 71-120. New York: Marcel Dekker Inc.

[47] Coombes, A. G. A., Rizzi, S. C., Williamson, M., Barralet J. E., Downes S., and Wallace, W. A. 2004. "Precipitation casting of poly-

caprolactone for applications in tissue engineering and drug delivery." *Biomat* 25: 315-325.

[48] Mondal, Debasish, Griffith, May., and Venkatraman, Subbu, S. 2016. "Polycaprolactone-based biomaterials for tissue engineering and drug delivery: Current scenario and challenges." *Int J Polym Mat Polym Biomat* 65:255-265.

[49] Wu, Chin-San. 2003. "Physical properties and biodegradability of maleated-polycaprolactone/starch composite." *Polym Deg Stab:* 80:127-134.

[50] Pitt, Collin C., Chasalow, F. I., Hibionada, Y. M., Klimas D. M., and Schindler, A. 1981. "Aliphatic polyesters. I. The degradation of poly (ε-caprolactone) *in vivo*." *J Appl Polym Sci* 81: 3779-3787.

[51] De Kesel, C., Vander, C., and Wauven, David C. 1997. "Biodegradation of polycaprolactone and its blends with poly(vinyl-alcohol) by micro-organisms from a compost of house-hold refuse." *Polym Deg Stabil* 55: 107-113.

[52] Labe, M., and Thielemans, W. 2009. "Synthesis of polycaprolactone: a review." *Chem Soc Rev* 38: 3484-504.

[53] Chun, Y. S., Kyung, Y. J., Jung, H. C., and Kim, W. N. 2000. "Thermal and rheological properties of poly(ε-caprolactone) and polystyrene blends." *Polym* 41: 8729-8733.

[54] Coleman, M. M., and Zarian J. 1979. "Fourier-transform infrared studies of polymer blends. II. Poly(ε-caprolactone)-poly(vinyl chloride) system." *J Polymer Science Polymer Physics* 17:837-850.

[55] Chen Biqiong, Sun Kang., and Ren Tao. 2005. "Mechanical and viscoelastic properties of chitin fiber reinforced poly(ε-caprolactone)." *Eur Polym J* 41:453-457.

[56] Honma Tomonori., Senda, Toshihiko, and Inoue Yoshio. 2003. "Thermal properties and crystallization behavior of blends of poly(ε-caprolactone) with chitin and chitosan." *Polym J* 52; 1839-1846.

[57] Hatakeyama, Hyoe, Iizuka, Yoshinobu., Kobashigawa Ken., Hirose Shigeo., and Hatakeyama Tatsuko. 1998. "Synthesis and physical properties of polyurethanes from saccharide-based polycaprolactones." *Macromol Symp* 130:127-138.

[58] Hatakeyama, Tatsuko., Izuta, Yoshinobu., Hirose, Shigeo., and Hatakeyama, Hyoe. 2002. "Phase transitions of lignin-based polycaprolactones and their polyurethane derivatives." *Polym* 43:177-1182.

[59] Hirose, Shigeo, Hatakeyama, Tatsuko., Izuta, Yoshinobu., and Hatakeyama, Hyoe. 2002. "TG-FTIR studies on lignin-based polycaprolactones." *J. Thermal Analysis and Calorimetry* 70: 853-860.

[60] Hatakeyama, Hyoe, Yoshida Takanori., and Hatakeyama, Tatsuko. 2000. "The Effect of side chain association on thermal and viscoelastic properties Cellulose acetate based polycaprolactones." *J Thermal Anal Cal* 59:157-168.

[61] Hatakeyama, Hyoe, Izuta, Yoshinobu., Kobashigawa, Ken., Hirose, Shigeo., and Hatakeyama, Tatsuko. 1998. "Synthesis and physical properties of polyurethanes from saccharide-based polycaprolactones." *Macromol Symp* 130:127-138.

[62] Hatakeyama, Hyoe., Nakano, Junzo., Hatano, Akira., and Migita, Nobuhiko. 1969. "Variation of infrared spectra with temperature for lignin and lignin model compounds." *Tappi* 52:1724-1728.

[63] Persenaire, Olivier, Alexandre Michaël., Degée, Philippe, and Dubois, Philippe. 2001. "Mechanisms and kinetics of thermal degradation of poly(ε-caprolactone)." *Biomacrom*ol 2: 288-294.

[64] Hirose, Shigeo, Hatakeyama, Tatsuko, Izuta, Yoshinobu, Hatakeyama, Hyoe. 2002. "TG-FTIR studies on lignin-based polycaprolactones." *J Therm Anal Cal* 70: 853-860.

[65] Hatakeyama, Tatsuko, and Hatakeyama, Hyoe. 1995. "Effect of chemical structure of amorphous polymers on heat capacity difference at glass transition temperature." *Thermochim Acta* 267:249-257.

[66] Chen B, Sun K, and Ren T. 2005. "Mechanical and viscoelastic properties of chitin fiber reinforced poly(ε-caprolactone)." *Eur Polym J* 41:453-457.

[67] Fabbri, Paola, Bondioli, Federica, Messori, Massimo, Bartoli, Cristina, Dinucci, Dinuccio, and Chiellini, Federica. 2010. "Porous scaffolds of polycaprolactone reinforced with in situ generated hydroxyapatite for bone tissue engineering." *J. Mat Sci. Mat Med.* 21: 343–351.

[68] Salgado, C. L., Sanchez, E. M. S., Mano, J. F., and Moraes, A. M. 2012. "Characterization of chitosan and polycaprolactone membranes designed for wound repair application." *J. Mat Sci* 47: 659–667.

[69] Lönnberg, Hanna, Karolina, Larsson., Tom Lindström., Anders Hult., and Eva Malmström. 2011. "Synthesis of polycaprolactone-grafted microfibrillated cellulose for use in novel bionanocomposites–influence of the graft length on the mechanical properties." *ACS Appl. Mater. Interfaces*, 3: 1426–1433.

In: Molasses: Forms, Production and Uses ISBN: 978-1-53614-703-2
Editors: K. Maddison and R. Fuller

Chapter 2

TUNISIAN CAROB MOLASSES (RUB EL KHARROUB): PROCESSING, USES AND CHARACTERISTICS

Leila Tounsi** and Nabil Kechaou
Research Group in Process Engineering Food,
National School of Engineers of Sfax,
University of Sfax, Sfax, Tunisia

ABSTRACT

This work contributes to the evaluation of the process technologies and the quality characteristics of the traditionally made carob molasses through a survey in some Tunisian regions (Monastir governorate). According to the questionnaires, carob molasses, known locally as 'Rub El Kharroub,' has been produced mainly by women using an artisanal process including practically manual operations carried out with domestic equipment. It has been used both in food, as a natural sweetener, and in folkloric medicine as a natural remedy. Four homemade carob molasses provided by different producers were examined for their physicochemical,

* Corresponding Author Email: leila.tounsi.enis@gmail.com.

nutritional, phytochemical and organoleptical properties. The characterization of the samples has revealed that the main physicochemical quality (mainly color and HMF concentration) were related to non- enzymatic browning reactions occurring during juice concentration. Moreover, they serve as a natural source of sugars, minerals and bioactive compounds endowed with biological effects (antioxidant and antibacterial activities). According to the presented results, this study encouraged the consumption of Tunisian carob molasses as a nutritious and healthy food and suggested its use as a functional ingredient in food and pharmaceutical industry.

Keywords: carob molasses, Tunisian survey, traditional process, physicochemical composition, phytochemical quality

INTRODUCTION

Molasses or syrup is a kind of fruit juice concentrate generally produced from different sugar-rich fruits for softening and conserving seasonal fruits. It is a traditional food well known in many Mediterranean countries as 'Rub' in Tunisia, 'Pekmez' in Turkey and 'Débès' in Lebanon (Dhaouadi et al., 2014; Tounsi et al., 2017) Fruit molasses is naturally nutritious food with high amounts of sugars, minerals and organic acids; therefore, it could be directly consumed or used as an ingredient in some food formulations such as ice cream, beverage and confectionery products (Karaca et al., 2012; Tounsi et al., 2017). In addition, it is a good source of phenolic compounds with health promoting properties such as antioxidant, antimicrobial and anticancer activities (Abbès et al., 2013; Dhaouadi et al., 2011, 2013, 2014; Wang et al., 2011). Molasses has been made with different techniques considering species of fruits used in production which consists mainly two stages of juice extraction and concentration (Sengül et al., 2007). Juice concentration as a heat treatment may promote chemical changes during molasses processing and storage. One of them is the non-enzymatic browning reactions, including mainly caramelization and Maillard reaction. Caramelization occurs by the sugar decomposition at high temperatures, while Maillard reaction takes place between amino acids and reducing sugars (Tounsi et al., 2017). Browning reactions are known to contribute to food quality and acceptability. They improve the sensory characteristics (color, aroma and flavor) of some food products such as coffee. They causes also loss of some nutrients (e.g., amino acids) and formation of other compounds such as brown pigment (melanoidins) and 5-hydroxymethyl-furfural (HMF) (Vaikousi et al., 2008). These products have been found to exhibited antioxidant and antibacterial activities, thus they could act as bioactive compounds (Benjakul et al., 2005; Rufián-Henares and Morales, 2007).

Carob pods are one of fruits commonly used in molasses production. They are the fruits of carob tree (*Ceratonia siliqua* L.) which is an evergreen tree cultivated or naturally grown mainly in Mediterranean countries, including Tunisia where it is distributed mostly along the coasts as a wild plant (Tounsi and Kechaou, 2017).

The production and consumption of carob molasses has been very widespread in Turkey, in this respect, several studies have been conducted on the physicochemical properties and nutritional composition of commercial products (Özcan et al., 2007; Sengül et al., 2007; Şimşek and Artik, 2002; Tetik et al., 2010, 2011; Toker et al., 2013). Carob Pekmez is a high energy food rich in sugars (mainly sucrose, glucose and fructose) and minerals (especially K, Ca, Mg, Na, P and Fe). Besides its nutritional compounds, it is characterized by brown to dark yellow colour However, to the best of our knowledge, little information is available concerning the production and characterization of carob molasses in Tunisia. According to the literature, only two research works have been carried out on the effect of processing on the chemical composition, functional properties and biological activities of samples made of Tunisian carob pods at lab scale (Dhaouadi et al., 2014; Tounsi et al., 2017).

This study is part of a large work aimed to highlight the characteristics of Tunisian carob molasses and promote its incorporation into food products. So that, the achievement of this main objective imposes a good identification and understanding of the processing conditions and a quality characterization of available samples in order to suggest strategies for improvement in the case of industrial application. Thus, the present work focused on describing the manufacturing technology of the traditional carob molasses, and evaluating the physicochemical, nutritional, phytochemical and sensory properties of some local products.

METHODS

Survey

The survey was conducted in some regions of Monastir governorate (Bekalta, Teboulba, Jammel and Khnis) (Figure 1) as an interview with 30 habitants (10 men and 20 women) via two questionnaires to get ideas about manufacturing steps and potential uses of carob molasses (Rub El Kharroub).

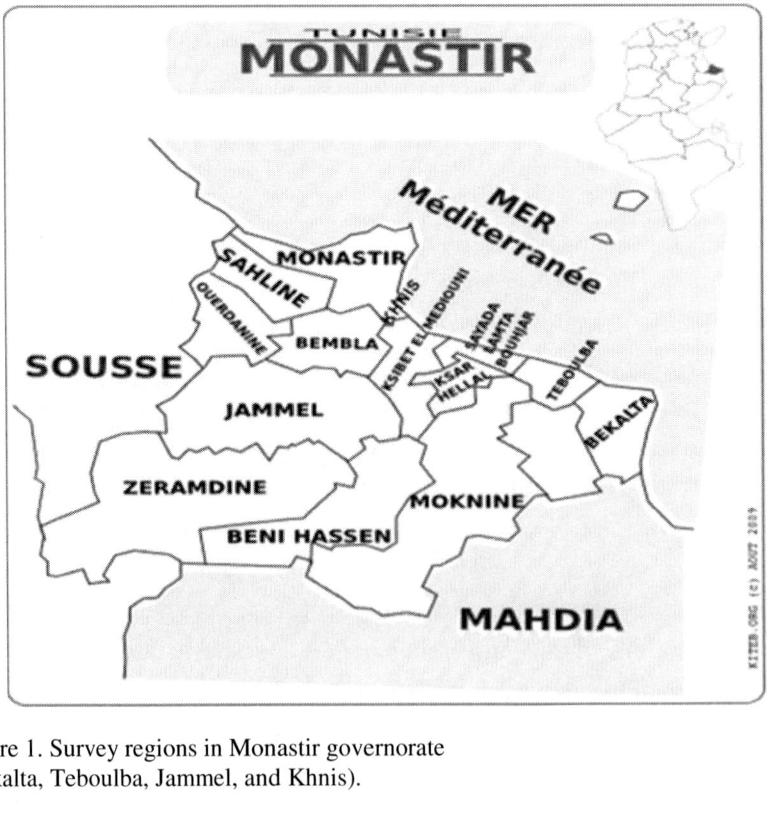

Figure 1. Survey regions in Monastir governorate (Bekalta, Teboulba, Jammel, and Khnis).

Samples

Four available homemade samples were kindly provided by different producers (one representative sample from each region) for the assay to determine their physicochemical, nutritional, phytochemical and organoleptical characteristics. Samples were kept in plastic containers at room temperature and marked by the first letter of corresponding regions as follows:

- Sample B: carob molasses coming from Bekalta region
- Sample T: carob molasses coming from Teboulba region
- Sample J: carob molasses coming from Jammel region
- Sample K: carob molasses coming from Khnis region

Physicochemical Analysis

Soluble solids content was determined using a hand refractometer (OpTech, Germany) and the results were expressed as Brix.

The color of carob molasses was determined using a colorimeter (Konica Minolta, Chroma Meter, CR 400-410, Japan). The calibration of the apparatus was done before the analysis using a white plate. Results were expressed as L* a* b* coordinates, where L* value ranges from 0 (black) to 100 (white); a* value ranges from negative (greenness) to positive (redness) and b* value ranges from negative (blueness) to positive (yellowness) according to CIE (1986).

The water activity (a_w) of molasses samples was analyzed at 25 °C by a a_w-meter apparatus (Sprint TH-500, Novasina, Switzerland).

The viscosity of carob molasses was directly measured at 25 °C with a viscosimeter (HA, Brookfield, USA) equipped with spindle 5 at the shear rate of 100 rpm and the results were expressed as mPa.s.

The pH of carob molasses was directly measured at 25 °C using a pH meter (METTLER TOLEDO MP 220, Switzerland).

The acidity of carob molasses was determined according the method described by AFNOR (1997). Briefly, the sample (1 g) was firstly dissolved in 100 ml of distilled water and then titrated with NaOH solution (0.1N) using phenolphthalein as an indicator. Results were expressed as g of citric acid (CA) per 100 g of molasses.

Browning index was measured as absorbance at 420 nm of diluted molasses (4 °Brix) according to Turkmen et al., (2006). Distilled water was used as a reference.

For HMF determination, molasses samples were firstly clarified following the method described by Toker et al., (2013) with some modifications. Briefly, each sample (500 mg) was suspended in 5 ml of distilled water, shaken vigorously for 1 min and mixed with 2 ml each of potassium ferrocyanide (15% w/v) and zinc acetate (30% w/v) solutions. After standing for 20 min, the mixture was filtered and the volume was made up to 50 ml with distilled water. The resulting filtrate was then assayed to quantify the HMF concentration according to Cohen et al., (1998). 1 ml of the filtrate was mixed thoroughly with 1 ml of trichloroacetic acid solution (TCA, 12% w/w) and 1 ml of thiobabituric acid solution (TBA, 0.025 M). The tubes with sample were then placed in the water bath at 40°C. After incubating for 50 min, the tubes were cooled immediately using tap water and the absorbance was measured at 443 nm. A calibration curve was performed from HMF solution within the range 0-10 mg/l.

Nutritional Analysis

Soluble sugars content was determined according to the phenol-sulfuric method (Dubois et al., 1956) using glucose solution as a standard (100 mg/l). Briefly, 200 μl of diluted molasses were mixed with 200 μl of phenolic solution (5%) and 1 ml of concentrated sulfuric acid (H_2SO_4). the mixture was allowed to stand for 30 min and the absorbance was then measured at 490 nm against the blank (distilled water instead of sample). Reducing sugars content was evaluated by the dinitrosalicylic acid (DNS) method using glucose solution (1 g/l) as a standard (Miller, 1959). The samples were first clarified as follows: 0.5 g of each sample was mixed with 10 ml of distilled water and 2 ml each of zinc acetate solution (30%) and potassium ferrocyanide solution (15%). After standing for 20 min, the mixture was filtered and the volume was made up to 50 ml with distilled water. 1 ml of the filtrate was mixed with 1 ml of DNS reagent in tubes that placed into a boiling water bath for 15 min. 10 ml of distilled water was added and the absorbance was measured at 575 nm against the blank (distilled water instead of sample). Sucrose content was estimated by calculating the difference between the contents of soluble and reducing sugars.

The content of proteins, lipids and minerals were determined according to AOAC (2000).

Proteins content was determined by the Kjeldahl method. The principle of this method is to measure the content of total nitrogen in the sample through three successive phases: digestion with concentrated sulfuric, distillation with sodium hydroxide solution and titration with acid solution. A conversion factor of 6.25 was used to convert the measured nitrogen content to proteins content.

Lipids content was determined by the Soxhlet extraction method. 5 g of each sample was extracted with hexane on a Soxhlet apparatus for 4 h. Solvent was then removed by vacuum evaporation and the residue of fat was weighed.

Minerals content was determined according to the dry ashing method. Samples (2 g) were incinerated in a muffle furnace at 550 °C until they become totally ash, and then they were weighed. In order to determine the minerals composition, an ash solution was prepared as described by Chew et al., (2011) with slight modifications. Resulting ash was dissolved in 20 ml hydrochloric acid (1 N). The mixture was then filtered in a 50 ml flask and the volume was made up with ultrapure water. Mineral concentrations (calcium, potassium, sodium, magnesium, iron, copper, zinc and manganese) were measured separately using an atomic absorption spectrophotometer (Thermo Scientific, ICE 3000, UK) according to AFNOR (1994). However, phosphorus concentration was analyzed colorimetrically by the molybdo-vanadate method (Kitson and Mellon, 1944).

Polyphenols content was determined according to the Folin–Ciocalteau method slightly modified (Singleton and Rossi, 1965). Beforehand, polyphenols were first extracted from carob molasses according to Avallone et al., (1997) with some modifications. 1 g of each sample was extracted twice with 10 ml of acetone 70% using a homogenizer at room temperature. After each centrifugation (10 min, 5000 rpm), the supernatants were pooled and filtered. 200 μl of diluted extract was mixed with 1 ml of Folin reagent (10 fold diluted) and 800 μl of saturated sodium carbonate solution (75 g/l). The mixture was placed in the dark for 15 min and then the absorbance was measured at 765 nm against the blank (acetone 70% instead of extract). A standard curve was prepared with gallic acid solutions ranged from 0 to 100 mg/l. Results are expressed as gallic acid equivalent (GAE).

Phytochemical Analysis

Antioxidant Activity

The antioxidant capacity of carob molasses samples (0.1% in distilled water) compared to the synthetic antioxidant BHT (0.1% in ethanol) was tested by three analytical methods:

- phosphomolybdenum assay to evaluate the total antioxidant activity. This assay is based on the reduction of Mo(VI) to Mo(V) in the presence of antioxidants and the subsequent formation of a green phosphate/Mo(V) complex at acidic pH (Prieto et al., 1999). 0.1 ml of diluted sample was mixed in an Eppendorf tube with 1 ml of reagent solution (0.6 M sulphuric acid, 28 mM sodium phosphate, and 4 mM ammonium molybdate). The tubes were incubated in water bath at 95 °C for 90 min. After cooling to room temperature, the absorbance was measured at 695 nm against the blank (distilled water instead of molasses sample). Results were expressed as equivalents of ascorbic acid (mg/g of sample).

- DPPH assay to evaluate the radical-scavenging activity. This assay is based on the decrease in absorbance at 517 nm of DPPH radical in the presence of antioxidants. Thus, lower absorbance of the reaction mixture indicated higher radical-scavenging activity (Bersuder et al., 1998). 500 μl of each sample solution was added to 375 μl of 99% ethanol and 125 μl of DPPH solution (0.02% in ethanol) as a free radical source. The mixture was shaken and then left to stand for 60 min in the dark at room temperature. Absorbance at 517 nm was measured using ethanol as blank. Results were expressed as percentage of DPPH inhibition and calculated as follows:

$$\text{DPPH inhibition (\%)} = [(A_{control} - A_{sample}) / A_{control}] \times 100$$

where A_{sample} is the absorbance of sample reaction and $A_{control}$ is the absorbance of the control reaction (containing all reagents except the sample).

- FRAP assay to evaluate the ferric reducing activity. This assay is based on the reduction of Iron (III) to iron (II) under acidic conditions in the presence of antioxidant. The ability of molasses to reduce iron was determined according to the method of Yildirim et al., (2001) with slight modifications. 0.5 ml of sample solution was mixed with 1.25 ml of potassium phosphate buffer (0.2M, pH 6.6) and 1.25 ml of 1% (w/v) potassium ferricyanide solution. The reaction mixtures were incubated in water bath at 50 °C for 20 min. After incubation, 0.5 ml of 10% trichloroacetic acid was added and the reaction mixture was then centrifuged at 2700 g for 10 min. Finally, 1.25 ml of the supernatant was mixed with 1.25 ml of distilled water and then 0.25 ml of 0.1% ferric chloride was added. After a 10 min reaction time, the absorbance was measured at 700 nm against the blank (distilled water instead of molasses sample). Results were expressed as absorbance values, and higher absorbance indicated higher reducing power activity.

Antibacterial Activity

The antimicrobial activity of the different samples was tested according to the Agar diffusion method described by Vanden Berghe and Vlietinck (1991). 200 µl of each bacterial culture were spread on LBA (Luria Bertani Agar). Once the plates had been aseptically dried, wells (7 mm depth, 6 mm diameter) were punched into the agar with a sterile borer and were loaded with 60 µl of molasses samples and Gentamicin antibiotic (50 mg/ml) as a positive control. The Petri dishes were kept first for 4 h at 4°C to allow sample diffusion in the agar, and then incubated for 24 h at 37°C. Antimicrobial activity was evaluated by measuring the diameter in millimeters (mm) of the inhibition zones around the well.

Ten bacterial strains, obtained from international culture collections (ATCC), were used as test microorganisms: *Listeria monocytogenes* (ATCC 43251), *Staphylococcus aureus* (ATCC 25923), *Escherichia coli* (ATCC 25922), *Salmonella enterica* (ATCC 43972), *Pseudomonas aeruginosa* (ATCC 27853), *Micrococcus luteus* (ATCC 4698), *Klebsiella pneumoniae* (ATCC 13883), *Enterococcus faecalis* (ATCC 29212), *Salmonella typhimurium* (ATCC 14028) and *Enterobacter* sp.

Organoleptical Analysis

An untrained panel from the students and the staff members of the National School of Engineers (Sfax, Tunisia) participated in the hedonic evaluation of carob molasses products sampled from survey locations (Monastir, Tunisia). The panel consisted of 44 consumers (11 men and 33 women) from 23 to 59 years. 4 codified samples were introduced to each panelist in a randomized way and they were evaluated for 6 descriptors (color, odor, viscosity, taste, aftertaste and overall acceptability) based on a hedonic scale from 1 (dislike extremely) to 9 (like extremely).

Statistical Analysis

All physicochemical analyses made on carob molasses samples were determined in triplicate and results were expressed as mean ± standard deviation. One-way analysis of variance (ANOVA) was used to compare the means by Duncan's multiple range test. Significant differences were detected at $P < 0.05$. Statistical analysis was made by SPSS software (Statistical Package for Social Science) version 20.0.

Results

Survey on Tunisian Carob Molasses

Carob molasses is a traditional product known as 'Rub El Kharroub' in Tunisia, mainly in Monastir governorate. Thus, the survey was carried out in some regions of Monastir (Bekalta, Teboulba, Jammel, and Khnis) using questionnaires to describe the processing steps as well as the potential uses.

Processing Steps

Studies of traditional products processing is very helpful for standardization of each step, minimization of producer painfulness, and amelioration of product yield and quality (Meli et al., 2013). For example, the manufacture of traditional date syrup in Algeria has been studied and improved by adapted equipment (Belguedj et al., 2015).

According to the questionnaires, only women, especially elderly in the range of 50-80 years, who master the manufacturing process of carob molasses. The analysis of the production technology shows a traditional process including practically manual operations carried out with limited equipment, mostly domestic utensils (Table 1).

Indeed, the general major steps in carob molasses manufacturing involved pretreatment of the raw material, water extraction of soluble solids and juice boiling until concentration. The principal raw material is the whole carob pods which may be either collected from farms or bought from the market. Carob pods were fragmented into large lumps or ground into small particles after washing and sun drying. The juice extraction was performed by mixing the treated pods with water for maceration and/or boiling, followed by filtration. Optionally, sucrose is supplemented to carob juice in the end of concentration to prepare more sweet syrup and reduce water activity for conservative purposes. The main procedure of manufacturing carob molasses reported in the present study was quite similar to that mentioned in a previous study for carob syrup preparation according to the traditional method used by Tunisian families (Dhaouadi et al., 2014). However, carob molasses in Turkey has been produced with different techniques including mainly the stages of water extraction, filtration/ clarification and concentration (Sengül et al., 2007; Tetik et al., 2010).

Table 1. Description of process steps of homemade carob molasses

Process steps	Operation	Equipments
Washing	Manual	• Large plastic containers • Tap water
Sun drying	Manual	Cloth covers
Grinding	Manual Mechanical	Mortar Hammer mill
Mixing	Manual	• Large plastic containers • Tap or rain water
Boiling	Manual	• Large metal vessels • Gaz cookers
Filtration	Manual	Sieve or filtering cloth
Packaging	Manual	Glass or plastic bottles

Figure 2 presented the common traditional process used by women in the survey locations (especially Bekalta region) for carob molasses production. Briefly, carob pods are washed, sun dried and ground manually or mechanically. They are then soaked in water for one night to induce soluble solids extraction. After filtration, the by-product is mixed again with water, and the pooled juice is boiled until concentration in an open vessel without sucrose supplementation. After cooling, the resulting juice concentrate is packaged and ready to be consumed or sold. The product thus prepared could be stored at room temperature for years (about 3-4 years) without any contamination as observed by the producing women and the consumers.

Regarding the by-product obtained after filtration, it is mostly considered as a solid waste; despite it still contains valuable compounds such as dietary fibers according to Papagiannopoulos et al., (2004). Recently, using by-products has become the focus of much research interest, supporting the economical biomass transformation. So that, the present study suggests the production of carob powder from the by-product of molasses processing in future works as a potential food valorization.

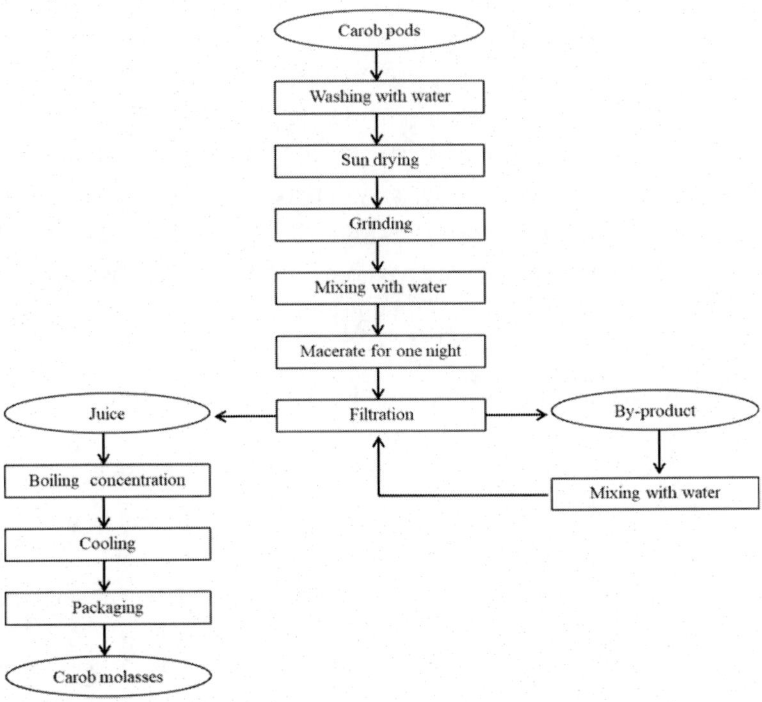

Figure 2. Common process diagram for carob molasses production in Monastir (Tunisia).

Potential Uses

Results of the survey conducted in the Monastir region revealed that traditional carob molasses, known locally as 'Rub El Kharroub,' is widely consumed by all the questioned people (10 men and 20 women) and used both in food and medicine.

On the one hand, it has been consumed mainly in breakfast as a natural sweetener along with other foodstuffs such as cooked dough, olive oil, bread and yoghurt. In the Maghreb countries, fruit molasses are also poured on cooked dough (traditional food known as 'assida') on specific occasions, such as the celebration of religious festivities (Dhaouadi et al., 2014). In Turkey, carob molasses, called 'carob Pekmez,' is widely consumed especially during the cold periods of the year as a source of energy. It is mainly consumed for breakfast instead of jam or marmalade or mixed with yoghurt (Atasoy, 2009; Karaca et al., 2012).

On the other hand, it has been also used in folkloric medicine as a natural remedy to treat some common diseases related to the digestive system (mouth to rectum) and the respiratory system (throat to lungs), as detailed in Table 2. These pharmacological actions of carob molasses are mainly attributed to the presence of bioactive compounds endowed with biological effects, including antioxidant, antimicrobial, anti-inflammatory and anti-tumor capacities (Dhaouadi et al., 2014; Wang et al., 2011). Carob Pekmez has been also used as antidiarrhoeal in the Aegean and Mediterranean parts of Turkey (Atasoy, 2009).

Table 2. Therapeutic effects of carob molasses according to the survey

Digestive system	Respiratory system
treating mouth ulcers	relieving sore throat
calming abdominal pain	calming the cough
relieving the hemorrhoid crisis	promoting expectoration
treating stomach disorders (ulcer, reflux ...)	treating common cold
treating intestines problems (constipation, diarrhea ...)	treating bronchitis

Characterization of Tunisian Carob Molasses

Representative samples of traditional carob molasses were collected from the four survey regions of Monastir governorate (Bekalta, Teboulba, Jammel, and Khnis) and then analyzed to evaluate their quality (physico-chemical, nutritional, phytochemical and sensorial properties). The four products were prepared following the main three processing steps (carob pods pretreatment, juice extraction and boiling concentration) with different methods as described in Table 3.

Table 3. Processing differences between the four homemade samples

Samples	Pretreatment	Extraction	Concentration
B	Manual grinding	Maceration	Without sugar addition
T	Mechanical grinding	Maceration	Without sugar addition
J	Fragmentation	Boiling	Without sugar addition
K	Fragmentation	Maceration, then boiling	With sugar addition

Physicochemical Quality

Table 4 presents the main physicochemical characteristics of homemade carob molasses sampled from the survey area (Monsatir, Tunisia).

Soluble solids content of the studied carob molasses (Rub El Kharroub) ranged from 75 to 80 °Brix, while that of commercial carob molasses in Turkey (carob Pekmez) did not exceed 75, ranging from 66.6 to 73.7 according to the literature (Sengül et al., 2007; Şimşek and Artik, 2002; Tetik et al., 2010, 2011; Toker et al., 2013). The Brix values of molasses were mainly due to their glucose, fructose and sucrose contents (Şimşek and Artik, 2002).

Table 4. Main physicochemical quality of traditional carob molasses samples

Molasses Samples	B	T	J	K
Soluble solids (Brix)	78 ± 0.00b	79.50 ± 0.50b	75 ± 0.02a	75.75 ± 0.35a
Water activity	0.75 ± 0.00b	0.77 ± 0.00b	0.66 ± 0.00a	0.69 ± 0.00a
Viscosity (mPa.s)	1916 ± 84.85b	2056 ± 67.88b	1284 ± 62.23a	1232 ± 11.31a
L*	30.73 ± 1.74a	31.92 ± 1.87a	31.28 ± 1.11a	30.32 ± 0.43a
a*	0.62 ± 0.10c	0.27 ± 0.04b	0.07 ± 0.00a	0.66 ± 0.01c
b*	-2.58 ± 0.21a	-2.57 ± 0.21a	-2.51 ± 0.08a	-2.24 ± 0.39a
pH	4.63 ± 0.07a	4.56 ± 0.00a	4.44 ± 0.00a	4.39 ± 0.01a
Acidity (g CA/100g)	1.45 ± 0.12a	1.71 ± 0.27a	1.37 ± 0.11a	1.54 ± 0.04a
Browning index	1.85 ± 0.01a	1.73 ± 0.03a	1.91 ± 0.06a	1.94 ± 0.01a
HMF (mg/100g)	36.46 ± 0.48b	31.31 ± 1.49a	31.98 ± 1.67a	38.66 ± 0.56c

Results are presented as mean ± standard deviation.

Values followed by different letters within the same line are statistically different ($P < 0.05$).

L* (100: white, 0: black), a* (+: red, -: green) and b* (+: yellow, -: blue).

The water activity values ranged from 0.66 to 0.77 for the four samples of carob molasses. No available data in the literature about the water activity of carob molasses, whereas, a previous study reported similar a_w values (about 0.7) for apricot and date molasses at 71.6 and 72.6 Brix, respectively (Karaman and Kayacier, 2011). Water activity is frequently mentioned as an important factor affecting non-enzymatic browning reaction rates. The rate of browning decreases at higher water activities (0.7–0.8), probably due to the increased dilution effect on solute concentration (Vaikousi et al., 2008). On the other hand, water activity is considered as a microbiological indicator. Indeed, the water activity values of the samples was lower than that required for the growth of some representative bacterial strains such as *Clostridium botulinum*, *Salmonella* sp., *Escherichia coli*, *Clostridium perfringens* and *Bacillus cereus*, which require water activities higher than 0.90, as well as for the growth of *Staphylococcus aureus*, which requires a minimum water activity of 0.86 (Guilherme et al., 2009).

Viscosity is considered as an important physical property related to the quality of liquid food products. The viscosity values of the studied carob molasses ranged between 1232 and 2056 mPa.s. It was observed from Table 4 that the highest viscosity values were attributed to the samples B and T which had the highest content of soluble solids. A previous study carried out in rheological characteristics of carob Pekmez at 72°Brix revealed lower viscosity values (~800 mPa.s at 30°C). In addition, it noted that carob molasses presented a non-newtonian thickening fluid and it has heterogeneous texture and non-uniform particle distribution (Sengül et al., 2007). Karaman and Kayacier (2011) stated also similar rheological behavior of apricot and date molasses as non-newtonian flow, with viscosity decreasing along the temperature. Guilherme et al., (2009) evaluated the quality of molasses prepared from mesquite pod and cashew apple and mentioned that their viscosity values were lower than those reported for honeys (4500-3600 mPa·s). The difference of the viscosity between molasses and honey might be attributed to their biochemical composition.

Regarding color parameters of Tunisian homemade carob molasses samples, L* values ranged from 30 to 32, a* values ranged from 0.07 to 0.66 and b* values from -2.24 to -2.58. These values seem to be in close proximity to those of black color (L* = 29.03, a* = -0.69, b* = -2.03) according to IEC (1999). In literature, several studies were conducted on colorimetric characterization of Turkish commercial carob molasses and different color values were reported. Indeed, the L*, a*, b* values were recorded as 19.35, 4.35, -6.01 according to Sengül et al., (2007); 18.28, 0.6, 0.51 according to Şimşek and Artik (2002); 16.33, 1.29, 1.21 according to Tetik et al., (2010) and 2.79, 1.31, 1.33 according to Toker et al., (2013). Accordingly, the studied carob molasses had high brightness (L*) and low redness (a*) values which indicate a good quality product. A high redness (a*) value is not desired because it occurs as a result of excessive browning reactions, mainly sugars caramelisation at high temperatures and Maillard reaction which takes place between amino acids and reducing sugars (Akbulut et al., 2008; Karaman and Kayacier, 2011; Sengül et al., 2007). On the other hand, Toker et al., (2013) studied the color properties of some molasses samples (grape, mulberry and carob molasses) and attributed the differences between the results to many factors including composition of fruits used as raw material, processing conditions (temperature and time), non-enzymatic browning reactions (caramelisation and Maillard reaction), pigments (carotenoids, anthocyanins and chlorophyll) degradation, ascorbic acid oxidation.

As can be seen from Table 4, pH and acidity values of the four collected samples were found to be close each other (4.4-4.7 and 1.37-1.71 g citric acid/100 g, respectively). Several studies determined the pH and acidity of commercial carob pekmez and reported different values in the range of 4.96-5.44 and 0.6-1.34 % as citric acid, respectively (Sengül et al., 2007; Şimşek and Artik, 2002; Tetik et al., 2010, 2011). In general, molasses were characterized by acidic pH which could be explained by either degradation of sugars into acids, or hydrolysis of HMF into levulinic and formic acid (Toker et al., 2013).

Molasses quality is often attributed to non-enzymatic browning reactions occurring during thermal processing of fruit juices (Tounsi et al., 2017). Several methods have been suggested for monitoring the extent of those reactions, including both color measurements and chemical analysis. Browning index (absorbance at 420 nm) has generally been used as a quick and easy method for brown pigment detection, while HMF concentration has been widely used as an indicator of excessive heating in foods containing sugars (Vaikousi et al., 2008). Browning index and HMF content of the four samples of Tunisian carob molasses were observed to be close each other in the range of 1.73-1.94 and 31.31-38.66 mg/100 g, respectively. Statistical analysis showed that the sample K had the highest HMF content; this result could be explained by the sugar supplementation during molasses processing which undergoes caramelisation reaction. Previous studies considered only HMF concentration as the main quality index of commercial carob pekmez and reported lower values ranged between 1.53 and 21.32 mg/kg (Özcan et al., 2007; Sengül et al., 2007; Şimşek and Artik, 2002; Tetik et al., 2010; Toker et al., 2013). Such differences between the molasses samples with respect to their HMF contents could be explained by several factors. One of them is the methods used for HMF determination (colorimetric assay or HPLC analysis). In addition, the composition of raw fruits, especially proteins and sugars which are substrates of non-enzymatic browning reactions, as well as phenolic compounds which may inhibit sometimes

Maillard reaction. Moreover, the effect of sugars type, for example fructose is about 40 times more reactive than glucose as a precursor to HMF formation. HMF formation in heat treated foods was also associated with processing temperature, pH and water activity of products, and concentration of divalent cations present in the environment (Toker et al., 2013).

As far as we know, compositional data on the traditional carob molasses in Tunisia (Rub El Khrroub) are provided for the first time. In conclusion, the main physicochemical characteristics of the studied carob molasses were related to non- enzymatic browning reactions occurring during juice concentration. Since colour and HMF content are the most important quality attributes for the acceptability of molasses, processing methods and conditions as well as storage conditions should be chosen carefully to minimize browning reactions.

Nutritional Composition

The proximate nutritional composition of the studied carob molasses (Rub El Kharroub) is given in Table 5.

Sugars were the predominant components in all the samples. Indeed, the soluble sugars content showed more than 50% for the four collected carob molasses, while the contents of reducing sugars and sucrose were found in the range of 22.85-25.6% and 26.02-33.63%, respectively. Similar findings were reported by other studies (Sengül et al., 2007; Şimşek and Artik, 2002; Tetik et al., 2010, 2011) for commercial products (carob Pekmez) which contained high amounts of total sugars (56.35-84.4%) composed mainly of invert sugars (17.05-49.10%) and sucrose (22.11-45.61%). A previous work determined the sugar profile of ten samples of commercial carob molasses in Turkey and indicated the sucrose as the main sugar followed by glucose and fructose in all samples (Tetik et al., 2011). The presence of such reducing sugars naturally in syrups could reduce their crystallization and provide a good source of rapid energy since they pass easily into the blood without digestion (Tounsi et al., 2017). It is noted also that the sample K had the highest contents of soluble sugars and sucrose according to ANOVA test; such result is mainly attributed to the sucrose supplemented to carob juice during boiling concentration to prepare the molasses.

Table 5. Proximate nutritional composition of traditional carob molasses samples

Molasses Samples	B	T	J	K
Soluble sugars*	54.46 ± 0.91b	51.09 ± 1.79a	50.81 ± 2.20a	61.37 ± 0.18c
Reducing sugars*	26.49 ± 1.22a	22.85 ± 2.12a	24.79 ± 3.87a	25.63 ± 3.26a
Sucrose*	27.97 ± 1.31a	28.24 ± 0.31a	26.02 ± 2.01a	33.63 ± 4.03b
Minerals*	1.34 ± 0.20a	1.57 ± 0.10a	1.46 ± 0.05a	1.59 ± 0.04a
Potassium**	885.65 ± 34.83a	987.84 ± 34.05a	893.27 ± 88.09a	990.36 ± 13.46a
Sodium**	224.52 ± 14.46a	232.44 ± 26.46a	268.85 ± 50.24a	286.54 ± 22.04a
Magnesium**	60.24 ± 7.58a	69.91 ± 3.29a	59.57 ± 4.23a	75.05 ± 4.18a
Calcium**	55.26 ± 4.94a	63.88 ± 8.85a	60.13 ± 4.66a	50.46 ± 5.54a
Phosphorus**	55 ± 1.30a	50.45 ± 4.2a	68.07 ± 1.11b	62.86 ± 5.10b
Iron**	3.28 ± 0.93a	3.41 ± 0.33a	3.32 ± 0.1a	3.03 ± 0.03a
Zinc**	1.89 ± 0.13a	1.71 ± 0.07a	1.97 ± 0.21a	2.6 ± 0.15b
Copper**	0.13 ± 0.01a	0.21 ± 0.01b	0.12 ± 0.01a	0.10 ± 0.01a
Manganese**	0.22 ± 0.01a	0.28 ± 0.04a	0.26 ± 0.03a	0.32 ± 0.02a
Polyphenols***	1.62 ± 0.05b	2.04 ± 0.04c	1.67 ± 0.02b	1.51 ± 0.02a
Proteins*	0.27 ± 0.01a	0.33 ± 0.13a	0.24 ± 0.04a	0.26 ± 0.03a
Lipids*	0	0	0	0

Results are presented as mean ± standard deviation.

Values followed by different letters within the same line are statistically different ($P < 0.05$).

*: g/100 g of molasses; **: mg/100 g of molasses; ***: g GAE/100 g of molasses.

The studied carob molasses were also found to contain minerals in the range of 1.3-1.6%. These values seem to be lower than those reported by other researchers for commercial carob Pekmez; 2.48% according to Sengül et al., (2007) and 2.16% according to Özcan et al., (2007). As shown in Table 5, the mineral composition of the samples was dominated by potassium and sodium (885.65-990.36 and 224.52-286.54 mg/100 g, respectively) followed by magnesium, calcium and phosphorus in the range of 50.45-75.05 mg/100

g. Considering the micronutrients, iron and zinc were the most abundant elements (~3 and ~2 mg/100 g, respectively), while copper and manganese were present at low level in the range of 0.1-0.3 mg/100 g. Previous studies conducted on the mineral profile of carob Pekmez have also stated potassium as major element with the highest concentration (1057.3-1607.47 mg/100 g). Besides, they have detected the other macroelements (Na, P, Ca and Mg) at different concentrations and the microelements (Cu, Fe, Zn and Mn) at very low concentrations (Özcan et al., 2007; Tetik et al., 2010). These differences found between mineral contents may arise from the differences between mineral composition of carob fruit used as raw material for molasses production or water used for juice extraction.

Regarding polyphenols, the phenolic content of the investigated carob molasses ranged from 1508.63 ± 20.29 to 2039.53 ± 41.54 mg GAE/100 g. The results found in the present study seem to be higher than those reported in the literature for some carob molasses products from different geographic locations, including Tunisia (699 ± 52 mg GAE/100 g; Dhaouadi et al., 2014 and 513,50-543,19 mg GAE/100 g dry matter; Tounsi et al., 2017), Turkey (716-1245 mg GAE/100 g dry matter; Tetik et al., 2011 and 162 ± 29 mg GAE/100 g dry matter; Tetik et al., 2010) and Germany (394 mg GAE/100 g; Papagiannopoulos et al., 2004). Variations in phenolic content could be probably attributed to the phenolic composition of carob fruit, or processing conditions of molasses production and even the methods used for polyphenolic extraction and estimation. In addition to phenolic content, the phenolic composition of carob molasses was determined and gallic acid was identified as the major compound with ~61% of total phenolics according to Dhaouadi et al., (2014) and ~93% of total phenolics according to Papagiannopoulos et al., (2004).

The studied carob molasses were found to contain very small content of proteins (0.24-0.33%). Sengül et al., (2007) reported similar result for carob Pekmez (0.33%) and explained this low concentration by the involvement of proteins in Maillard reaction occurring during the heat treatment. Moreover, the four samples of carob molasses did not present any oil drops (Table 5). A previous study (Özcan et al., 2007) conducted on the compositional properties of carob pekmez did not also detected the presence of oil normally as the fruit had very low amounts of lipids in the range of 0.23%.

Summing up the results, it can be conclude that carob molasses is rich in sugars, devoid of oil and contains appreciable amounts of minerals and very low levels of proteins. Similar nutritional composition was reported for molasses produced from others fruits such as date (Abbès et al., 2011) and juniper (Akbulut et al., 2008). In this respect, thanks to high contents of sugars and minerals, molasses could be considered as a very important food product in human nutrition, especially for babies, children, sportsmen, and pregnant women (Akbulut et al., 2008; Sengül et al., 2007; Şimşek and Artik, 2002). Furthermore, the high sugar content could justify their use as a sweetening and flavoring agent in many food products such as beverages, ice cream, pastry products, confectionery products and dairy products (Abbès et al., 2011). Beside nutritional components, carob molasses were also found to be a good source of phenolic compounds with health promoting effects as well as various biological activities, mainly antioxidant, antimicrobial and cytotoxic potentials (Dhaouadi et al., 2014). That is one of the reasons why this processed product should be used as a functional food.

Phytochemical Properties

Phytochemicals are non-nutritive plant substances possessing varying degrees of disease-preventive properties. Generally, they are bioactive compounds such as phenolics, flavonoids, vitamins, and essential oils which are known to exhibit certain health benefits including antioxidative, anti-inflammatory, antitumor, and antimicrobial activities (Oikeh et al., 2016). In this study, the chemical composition of Tunisian carob molasses revealed the presence of bioactive compounds, including HMF (Table 4) and polyphenols (Table 5) which justify the evaluation of their phytochemical properties, namely antioxidant and antimicrobial activities.

Antioxidant Activity

In order to evaluate the antioxidant activity of a natural product, it is crucial to perform more than one antioxidant test, taking into consideration the various oxidation reactions. In this study, three assays, including total antioxidant activity, radical-scavenging activity and ferric reducing activity, were employed to evaluate the antioxidant capacity of the four collected samples of carob molasses compared to the synthetic antioxidant BHT at the same concentration (0.1%) (Figure 3).

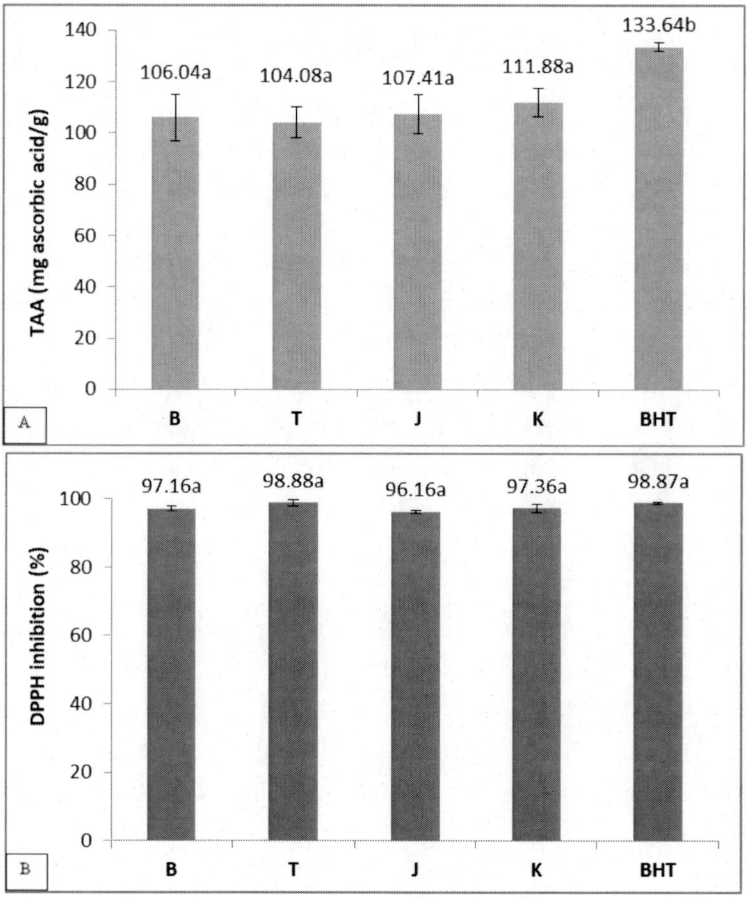

Figure 3. Continued.

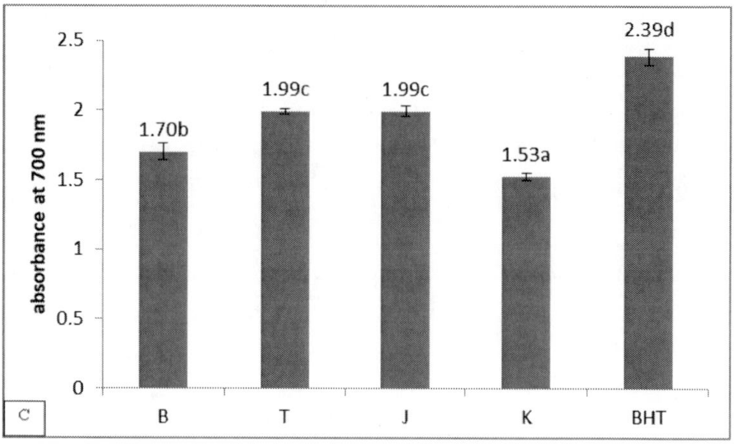

Figure 3. Antioxidant activities of carob molasses samples (B, T, J, K) compared to the reference antioxidant BHT: (A) total antioxidant activity, (B) radical-scavenging activity, (C) ferric reducing activity. Means marked by different letters are statistically different ($P < 0.05$).

The total antioxidant activity (TAA) of the various samples was measured using the phosphomolybdenum method which is based on the reduction of Mo(VI) to Mo(V) by the antioxidant compound and the formation of a green phosphate/Mo(V) complex (Prieto et al., 1999). As shown in Figure 3A, the four collected samples exhibited important TAA (~104-112 mg ascorbic acid/g carob molasses) if compared with the positive control (133.64 mg ascorbic acid/g BHT).

The antioxidant potential of carob molasses samples was also investigated by DPPH test which is widely used to determine the antiradical activity of antioxidant compounds. DPPH as a free radical shows maximum absorbance at 517 nm, whereas in the presence of antioxidants donors of electrons, the radical would be scavenged and the absorbance is reduced (Bersuder et al., 1998). Figure 3B clearly indicated that the samples of homemade carob molasses were found capable to reduce the DPPH radical as well as the BHT antioxidant with DPPH inhibition in the range of 96-99%.

Moreover, the antioxidant capacity of the Tunisian carob molasses was evaluated using the FRAP assay often used to test the ability of an antioxidant compound to reduce Fe^{3+} to Fe^{2+} by donating electrons (Yildirim et al., 2001). The reducing power of the molasses samples compared with BHT is presented in Figure 3C as absorbance at 700 nm. It is known that the higher the absorbance is, the greater is the reducing power. It was observed that the studied samples showed a reducing activity closer to that of the reference antioxidant (1.5-2 vs 2.4, respectively). The reducing properties are generally associated with the presence of reductones, which have been shown to exert an antioxidant action by donating a hydrogen atom, thus breaking the free radical chains and converting them into more stable (Abbès et al., 2013).

For all the antioxidant evaluation tests, the results showed that the four samples of homemade carob molasses exhibited interesting antioxidant potentials as well as the synthetic antioxidant BHT. These findings are in agreement with those reported by Dhaouadi et al., (2014) concerning the antiradical activities of traditional carob syrup which were higher than those of the well-known antioxidant, ascorbic acid. The authors attributed the strong antioxidant properties mainly to the high content of polyphenolic compounds capable to reduce free radicals due to their hydrogen donating ability. Other studies conducted on Tunisian traditionally made syrups prepared from date and barbary-fig indicated that phenolic compounds and flavonoids are the components responsible for the antioxidant effects of these syrups (Dhaouadi et al., 2011, 2013). To explain more and more the antioxidant effect of molasses, previous works carried out in molasses derived from different plant materials including carob, cane, grape, sorghum (Wang et al., 2011) and date (Abbès et al., 2013) reported positive and high correlation coefficients between the polyphenol levels and the antioxidant activity determined by several methods. They also suggested other antioxidant compounds such as peptides, organic acids and Maillard reaction products (MRPs) which are naturally produced in food during thermal processing and storage by interaction between reducing sugars and available amino acids.

In fact, non-enzymatic reactions products could act as naturally produced antioxidants in fruit juice processing. MRPs, especially melanoidins, as well as caramelisation products (CPs), mainly HMF compound, have been found to exhibit antioxidative activities such as radical scavenging activity, metal chelating activity, reducing power, scavenging of active oxygen species and decomposition of hydroperoxide (Benjakul et al., 2005; Chen et al., 2009; Hwang et al., 2011; Rufián-Henares and Morales, 2007; Turkmen et al., 2006). Another study have also evaluated the radical scavenging activity and the ferric-reducing power of different citrus juice concentrates, and attributed their antioxidant activities to the presence of phytochemical compounds, mainly flavonoids, alkaloids, steroids, terpenoids, saponins, cardiac glycosides, and reducing sugars (Oikeh et al., 2016).

Antibacterial Activity

The antibacterial activity of the collected samples was evaluated by the agar well diffusion assay against ten different bacterial strains which are known as opportunistic human and animal pathogens and/or cause food contamination and quality deterioration. The obtained results of growth inhibition for these strains are shown in Table 6.

Table 6. Antibacterial activity, expressed as inhibition zone diameter (mm), of carob molasses samples compared to the reference antibiotic Gentamicin

Molasses samples	B	T	J	K	Gentamicin
Enterobacter sp.	25.0 ± 0.0b	20.0 ± 0.0a	19.0 ± 1.0a	21.0 ± 1.0a	27.0 ± 0.0c
E. coli	27.5 ± 2.5c	21.0 ± 1.0b	20.5 ± 0.5b	16.5 ± 0.5a	24.5 ± 0.5c
E. faecalis	15.0 ± 0.0a	20.0 ± 0.0b	29.0 ± 1.0c	22.5 ± 2.5b	21.0 ± 1.0b
K. pneumonia	22.5 ± 2.5b	25.0 ± 0.0c	14.0 ± 1.0a	17.5 ± 2.5a	25.5 ± 0.5c
L. monocytogenes	29.5 ± 0.5b	24.5 ± 0.5a	26.0 ± 1.0a	31.0 ± 1.0b	24.0 ± 1.0a
M. luteus	33.5 ± 1.5c	26.5 ± 1.5b	30.0 ± 2.0c	20.5 ± 0.5a	24.5 ± 0.5b
P. aeruginosa	15.0 ± 1.0a	19.5 ± 0.5b	30.5 ± 0.5c	17.0 ± 1.0a	21.0 ± 1.0b
S. aureus	27.5 ± 2.5d	15.0 ± 0.0a	19.0 ± 1.0b	15.5 ± 0.5a	23.0 ± 1.0c
S. enterica	25.0 ± 0.0b	24.5 ± 0.5b	14.5 ± 1.5a	18.0 ± 2.0a	27.5 ± 0.5c
S. typhimurium	25.0 ± 0.0b	20.0 ± 0.0a	20.5 ± 0.5a	31.0 ± 1.0c	30.0 ± 0.0c

Results are presented as mean ± standard deviation.

Values followed by different letters within the same line are statistically different ($P < 0.05$).

It was observed that the Tunisian homemade carob molasses exhibited interesting antimicrobial potential towards all the tested microorganisms, in comparison with the reference antibiotic Gentamicin. Indeed, the four molasses products were able to inhibit the growth of the bacterial germs with various degrees and to generate inhibition zones even more important than the synthetic antibiotic, especially against Listeria monocytogenes, Micro-coccus luteus and Pseudomonas aeruginosa (Table 6). Carob molasses samples had also an inhibitory effect against Staphylococcus aureus which is well-known for its resistance to some phytochemical compounds and for the production of several types of enterotoxins that cause gastroenteritis (Halpin-Dohnalek and Marth, 1989).

According to Dhaouadi et al., (2014), carob syrup traditionally processed from Tunisian carob pods showed also an important growth inhibition against eight different bacterial strains: *Staphylococcus aureus*, *Staphylococcus epidermidis*, *Bacillus cereus*, *Bacillus subtilis*, *Strepto-coccus faecalis*, *Pseudomonas aeruginosa*, *Escherichia coli* and *Salmonella sp*. The authors reported this strong antimicrobial activity to the presence of high content of polyphenols in carob syrup extract, mainly gallic acid and epigallocatechin gallate.

Other studies carried out in the antibacterial activity of fruit molasses (date and barbary-fig molasses) attributed the inhibitory effect against bacterial pathogens to their phenolic composition. The bacteriostatic and/ or bactericide action of these kind of compounds could be explained by their ability to adsorb to cell membrane and to disrupt its permeability barrier, through interaction with enzymes and effectors, or deprivation of substrates and metal ions (Dhaouadi et al., 2011, 2013). In addition to phenolics, other bioactive compounds formed during molasses processing could contribute to their biological activities. Indeed, the products of non-enzymatic browning reactions (melanoidins and HMF) were found to exert an antibacterial effect by binding essential metals or precipitating proteins (Rufián-Henares and Morales, 2007; Wang et al., 2011). Previous study have attributed antibacterial activities of different citrus juice concentrates to the presence of phytochemical compounds, mainly flavonoids, alkaloids, steroids, terpenoids, saponins, cardiac glycosides, and reducing sugars (Oikeh et al., 2016).

Based on the above results, it can be concluded that the research into the phytochemical properties of the homemade carob molasses has been very successful. Indeed, the four products exhibit important biological effects (antioxidant and antibacterial activities) closely attributed to their polyphenolic constituents and other bioactive compounds, such as melanoidins and HMF, products of non-enzymatic browning reactions. Thus, Tunisian traditional carob molasses (Rub El Kharroub) may be considered as a natural source of phytochemical compounds which justify their uses in traditional medicine.

Hedonic Evaluation

Hedonic evaluation was conducted to study the consumers' preference of the four samples of carob molasses and the mean scores of the sensory analysis are shown in Figure 4.

Concerning color and odor, there is no significant difference ($P > 0.05$) between the four samples according to the statistical analysis (ANOVA test). They presented sensory scores in the range of 4-5 corresponding to "dislike slightly."

Concerning viscosity, consumers gave the highest scores (~6) for the samples J and K which had the lowest viscosity values (Table 4).

Concerning taste and aftertaste, consumers gave the highest scores (~6) for the sample K. this result may be explained by the sweet taste due to the sucrose addition during the product processing.

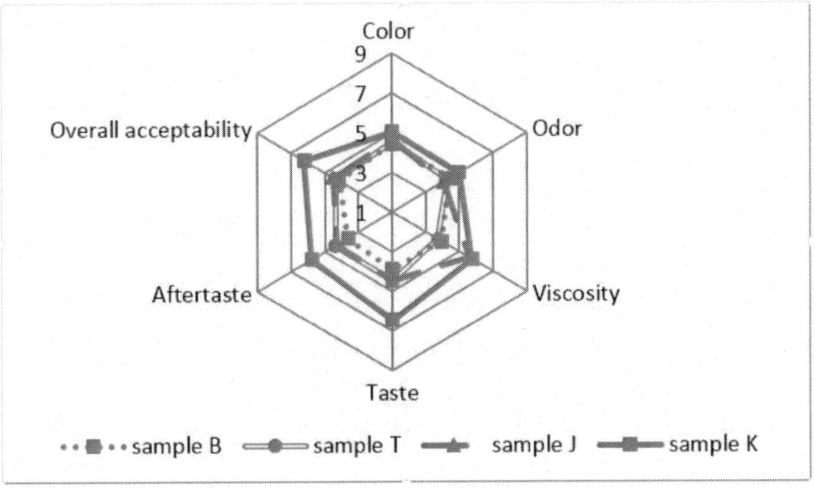

Figure 4. Hedonic scores of the four samples of homemade carob molasses from Tunisia: 1 = dislike extremely; 2 = dislike very much; 3 = dislike moderately; 4 = dislike slightly; 5 = neither like nor dislike; 6 = like slightly; 7 = like moderately; 8 = like very much; 9 = like extremely.

Concerning overall acceptability, the sample K was the most appreciated by the consumers as it presented the highest score (6.23) corresponding to "like slightly." In this respect, it could be conclude that sugar supplementation during traditional carob molasses processing improved its sensorial quality, mainly taste. However, it affected its nutritional composition and biological activities; and thus it caused health negative attributes associated with physiological disorders (mainly, diabetes and obesity) according to a previous study done by Dhaouadi et al., (2014). For the other samples processed without sugar addition (samples B, T and J), they presented mean scores in the range of rejection (~4) corresponding to "dislike slightly."

As conclusion, the studied samples, excluding the sample K, were not almost overall accepted by consumers; this result was expected since most of consumers have never consumed carob molasses. Indeed, carob fruit and its products such as syrup and powder, as well as their consumption are not well spread all over the country (Tunisia), which might have contributed to

a less acceptance of the molasses product. On the other hand, the hedonic results found in the present study could be due the fact that carob molasses had dark color and high viscosity; especially that color and viscosity are the most important quality attributes which affect consumer acceptability of foodstuffs.

No published data on the sensory quality of carob molasses were found in the literature; hence this study describes for the first time the consumers' preference of carob molasses traditionally made by Tunisian women. However, our findings could be compared to those reported by a previous work conducted on quality evaluation of molasses made of other fruits from Brazil. The authors showed that cashew apple syrup was more appreciated than mesquite pod syrup which might be also attributed to consumers' habits and products commercialization (Guilherme et al., 2009).

CONCLUSION

This study describes for the first time the processing and the properties of the homemade carob molasses in Tunisia known as 'Rub El Kharroub.'

According to the survey, this local product has been generally prepared with a tradition process by Tunisian families in some coastal regions, mainly Monastir. The traditionally made carob molasses exhibited different properties which justify its consumption as a nutritional and functional food. Indeed, it has been widely consumed as an energetic food or as a natural sweetener for its high sugar content. It has been also used in folkloric medicine as a natural remedy thanks to its bioactive compounds (phenolic substances and products of non-enzymatic browning reactions) and thus their biological effects (antioxidant and antibacterial activities).

Regarding its nutritional, organoleptical and phytochemical quality, Tunisian carob molasses could be also exploited in food industry and used as natural ingredient or additive. In this respect, this study suggested its incorporation in foodstuffs (e.g., confectionery products), in future works.

REFERENCES

Abbès, F., Bouaziz, M. A., Blecker, C., Masmoudi, M., Attia, H., and Besbes, S. 2011. "Date syrup: Effect of hydrolytic enzymes (pectinase/ cellulase) on physico-chemical characteristics, sensory and functional properties." *LWT - Food Science and Technology* 44:1827–1834.

Abbès, F., Kchaou, W., Blecker, C., Ongena, M., Lognay, G., Attia, H., and Besbes, S. 2013. "Effect of processing conditions on phenolic compounds and antioxidant properties of date syrup." *Industrial Crops and Products* 44:634–642.

AFNOR. 1994. *Jus de fruits et de légumes - Dosage des minéraux par spectrométrie d'absorption atomique. NF V76-117.* Paris, France : Association Française de Normalisation. *[Fruit and vegetable juices - Determination of minerals by atomic absorption spectrometry. NF V76-117.* Paris, France: French Association of Standardization.]

AFNOR. 1997. *Jus de fruits et de légumes - Détermination de l'acidité titrable. NF 12147.* Paris, France : Association Française de Normalisation. [*Fruit and vegetable juices - Determination of titratable acidity. NF 12147.* Paris, France: French Association of Standardization.]

Akbulut, M., Coklar, H., and Ozen, G. 2008. "Rheological Characteristics of Juniperus drupacea Fruit Juice (pekmez) Concentrated by Boiling." *Food Science and Technology International* 14:321–328.

AOAC. 2000. *Official methods of analysis (17th ed).* Washington, USA: Association of Official Analytical Chemists.

Atasoy, A. F. 2009. "The effects of carob juice concentrates on the properties of yoghurt." *International Journal of Dairy Technology* 62:228–233.

Avallone, R., Plessi, M., Baraldi, M., and Monzani, A. 1997. "Determination of Chemical Composition of Carob (Ceratonia siliqua): Protein, Fat, Carbohydrates, and Tannins." *Journal of Food Composition and Analysis* 10:166–172.

Belguedj, N., Bassi, N., Fadlaoui, S., Agli, A., Constantine, M. De, and Constantine, M. De. 2015. 'Contribution à l'industrialisation par l'amélioration du processus traditionnel de fabrication de la boisson locale à base de datte'. *Journal of New Sciences* 20:818–829. ["Contribution to industrialization by improving the traditional process of making the local date drink." *Journal of New Sciences* 20:818–829.]

Benjakul, S., Visessanguan, W., Phongkanpai, V., and Tanaka, M. 2005. "Antioxidative activity of caramelisation products and their preventive effect on lipid oxidation in fish mince." *Food Chemistry* 90:231–239.

Bersuder, P., Hole, M., and Smith, G. 1998. "Antioxidants from a heated histidine-glucose model system. I: Investigation of the antioxidant role of histidine and isolation of antioxidants by high-performance liquid chromatography." *Journal of the American Oil Chemists' Society* 75:181–187.

Chen, S. L., Yang, D. J., Chen, H. Y., and Liu, S. C. 2009. "Effect of hot acidic fructose solution on caramelisation intermediates including colour, hydroxymethylfurfural and antioxidative activity changes." *Food Chemistry* 114:582–588.

Chew, L. Y., Prasad, K. N., Amin, I., Azrina, A., and Lau, C. Y. 2011. "Nutritional composition and antioxidant properties of *Canarium odontophyllum* Miq. (dabai) fruits." *Journal of Food Composition and Analysis* 24:670–677.

CIE. 1986. *Colorimetry (Supplement N°2, Publication N°15)*. Paris, France: Commision International de l'Eclairage. [*Colorimetry (Supplement N°2, Publication N°15)*. Paris, France: International Lighting Commission].

Cohen, E., Birk, Y., Mannheim, C., and Saguy, I. 1998. "A Rapid Method to Monitor Quality of Apple Juice During Thermal Processing." *LWT - Food Science and Technology* 31: 612–616.

Dhaouadi, K., Belkhir, M., Akinocho, I., Raboudi, F., Pamies, D., Barrajón, E., and Fattouch, S. 2014. "Sucrose supplementation during traditional carob syrup processing affected its chemical characteristics and biological activities." *LWT - Food Science and Technology* 57:1–8.

Dhaouadi, K., Raboudi, F., Estevan, C., Barrajon, E., Vilanova, E., Hamdaoui, M., and Fattouch, S. 2011. "Cell viability effects and antioxidant and antimicrobial activities of Tunisian date syrup (Rub El Tamer) polyphenolic extracts." *Journal of Agricultural and Food Chemistry* 59:402–406.

Dhaouadi, K., Raboudi, F., Funez-Gomez, L., Pamies, D., Estevan, C., Hamdaoui, M., and Fattouch, S. 2013. "Polyphenolic Extract of Barbary-Fig (Opuntia ficus-indica) Syrup: RP-HPLC-ESI-MS Analysis and Determination of Antioxidant, Antimicrobial and Cancer-Cells Cytotoxic Potentials." *Food Analytical Methods* 6:45–53.

Dubois, M., Gilles, K. A., Hamilton, J. K., Rebers, P. A., and Smith, F. 1956. "Colorimetric Method for Determination of Sugars and Related Substances." *Analytical Chemistry* 28:350–356.

Guilherme, A. A., Honorato, T. L., Dornelles, A. S., Pinto, G. A. S., Brito, E. S., and Rodrigues, S. 2009. "Quality evaluation of mesquite (*Prosopis juliflora*) pods and cashew (*Anacardium occidentale)* apple syrups." *Journal of Food Process Engineering* 32:606–622.

Halpin-Dohnalek, M. I., and Marth, E. H. 1989. "*Staphylococcus aureus* : Production of Extracellular Compounds and Behavior in Foods - A Review." *Journal of Food Protection* 52:267–282.

Hwang, I. G., Kim, H. Y., Woo, K. S., Lee, J., and Jeong, H. S. 2011. "Biological activities of Maillard reaction products (MRPs) in a sugar-amino acid model system." *Food Chemistry* 126: 221–227.

IEC. 1999. *Multimedia systems and equipment - Colour measurement and management (International Standard 61966-2-1)*. Geneva, Switzerland: International Electrotechnical Commission.

Karaca, O. B., Saydam, I. B., and Güven, M. 2012. "Physicochemical, mineral and sensory properties of set-type yoghurts produced by addition of grape, mulberry and carob molasses (Pekmez) at different ratios." *International Journal of Dairy Technology* 65:111–117.

Karaman, S., and Kayacier, A. 2011. "Effect of temperature on rheological characteristics of molasses: Modeling of apparent viscosity using Adaptive Neuro - Fuzzy Inference System (ANFIS)." *LWT - Food Science and Technology* 44:1717–1725.

Kitson, R. E., and Mellon, M. G. 1944. "Colorimetric Determination of Phosphorus as Molybdivanadophosphoric Acid." *Industrial and Engineering Chemistry Analytical Edition*, 16:379–383.

Meli, C. F., Njintang, N. Y., Noumi, G. B., Bernard, C., Relkin, P., Armand, M., and Mbofung, C. M. F. 2013. "Processing, Physicochemical,

Nutritional, and Organoleptical Properties of Dackere, an African Cereal/Tuber Food Semolina." *Food and Bioprocess Technology* 6:699–709.

Miller, G. L. 1959. "Use of Dinitrosalicylic Acid Reagent for Determination of Reducing Sugar." *Analytical Chemistry* 31:426–428.

Oikeh, E. I., Omoregie, E. S., Oviasogie, F. E., and Oriakhi, K. 2016. "Phytochemical, antimicrobial, and antioxidant activities of different citrus juice concentrates." *Food Science and Nutrition* 4:103–109.

Özcan, M. M., Arslan, D., and Gökçalik, H. 2007. "Some compositional properties and mineral contents of carob (*Ceratonia siliqua*) fruit, flour and syrup." *International Journal of Food Sciences and Nutrition* 58:652–658.

Papagiannopoulos, Menelaos, Wollseifen, H. R., Mellenthin, A., Haber, B., and Galensa, R. 2004. "Identification and quantification of polyphenols in carob fruits (*Ceratonia siliqua*) and derived products by HPLC-UV-ESI/MS." *Journal of Agricultural and Food Chemistry* 52:3784–3791.

Prieto, P., Pineda, M., and Aguilar, M. 1999. "Spectrophotometric Quantitation of Antioxidant Capacity through the Formation of a Phosphomolybdenum Complex: Specific Application to the Determi-nation of Vitamin E." *Analytical Biochemistry* 269:337–341.

Rufián-Henares, J. A., and Morales, F. J. 2007. "Functional properties of melanoidins: In vitro antioxidant, antimicrobial and antihypertensive activities." *Food Research International* 40:995–1002.

Sengül, M., Ertugay, M. F., Sengül, M., and Yüksel, Y. 2007. "Rheological characteristics of carob Pekmez". *International Journal of Food Properties* 10:39–46.

Şimşek, A., and Artik, N. 2002. "Studies of composition of concentrates from different fruit." *GIDA* 27:459–467.

Singleton, V. L., and Joseph A. Rossi, J. 1965. "Colorimetry of total phenolics with phosphomolybdic-phosphotungstic acid reagents." *American Journal of Enology Viticulture* 16:144–158.

Tetik, N., Turhan, İ., Karhan, M., and Öziyci, H. R. 2010. "Characterization of, and 5-hydroxymethylfurfural concentration in carob Pekmez." *GIDA* 35:417–422.

Tetik, N., Turhan, I., Oziyci, H. R., and Karhan, M. 2011. "Determination of D-pinitol in carob syrup." *International Journal of Food Sciences and Nutrition* 62:572–576.

Toker, O. S., Dogan, M., Ersöz, N. B., and Yilmaz, M. T. 2013. "Optimization of the content of 5-hydroxymethylfurfural (HMF) formed in some molasses types : HPLC-DAD analysis to determine effect of different storage time and temperature levels." *Industrial Crops and Products* 50:137–144.

Tounsi, L., Karra, S., Kechaou, H., and Kechaou, N. 2017. "Processing, physico-chemical and functional properties of carob molasses and powders." *Journal of Food Measurement and Characterization* 11:1440–1448.

Tounsi, L., and Kechaou, N. 2017. *Le caroubier (Ceratonia siliqua L.) et ses fruits: descriptions, intérêts et applications.* Sarrebruck, Allemagne : Éditions universitaires européennes. [*The carob tree (Ceratonia siliqua L.) and its fruits: descriptions, interests and applications*. Saarbrücken, Germany: European Academic Publishing].

Turkmen, N., Sari, F., Poyrazoglu, E. S., and Velioglu, Y. S. 2006. "Effects of prolonged heating on antioxidant activity and colour of honey." *Food Chemistry* 95:653–657.

Vaikousi, H., Koutsoumanis, K., and Biliaderis, C. G. 2008. "Kinetic modelling of non-enzymatic browning of apple juice concentrates differing in water activity under isothermal and dynamic heating conditions." *Food Chemistry* 107:785–796.

Vanden Berghe, D. A., and lietinck, A. J. 1991. "Screening methods for antibacterial and antiviral agents from higher plants." In, *Methods in Plant Biochemistry.* Hostettmann, K. (Ed.), Academic Press, London, UK.

Wang, B. Sen, Chang, L. W., Kang, Z. C., Chu, H. L., Tai, H. M., and Huang, M. H. 2011. "Inhibitory effects of molasses on mutation and nitric oxide production." *Food Chemistry* 126:1102–1107.

Yildirim, A., Mavi, A., and Kara, A. A. 2001. "Determination of Antioxidant and Antimicrobial Activities of *Rumex crispus* L. Extracts." *Journal of Agricultural and Food Chemistry* 49:4083–4089.

In: Molasses: Forms, Production and Uses ISBN: 978-1-53614-703-2
Editors: K. Maddison and R. Fuller

Chapter 3

Molasses: Desugarization Processes and Purification Treatments

Miljana Djordjević*[*], *Zita Šereš, Nikola Maravić and Marijana Djordjević
Department of Carbohydrate Food Engineering,
Faculty of Technology Novi Sad,
University of Novi Sad, Novi Sad, Serbia

Abstract

Considering that molasses is produced at about 2 – 5% on the starting raw material (sugar beet or sugar cane) depending on the raw material condition and applied processing operations, considerable amounts of sucrose can be recovered and an increase in the efficiency of the sugar factory achieved. Molasses potential in this field was recognized since sucrose content alongside other sugars accounts for nearly 50%. The presented chapter provides a comprehensive review on sucrose recovery from molasses referred as molasses desugarization process by disclosing common and advanced technologies applied in industry alongside ongoing research in this field. Attention was also directed towards the main results

[*] Corresponding Author Email: miljanadj@tf.uns.ac.rs.

issued from studies on molasses purification by means of standard and emerging filtration aids utilization and application of membrane separation processes as well as combination of these treatments. Particular emphasis was given to recent introduction of bentonite as an adsorbent of molasses non-sugars with the focus on molasses quality enhancement and determination of optimal treatment conditions. A new approach for molasses purification which combines the use of an adsorbent and membrane separation process is also discussed.

Keywords: molasses, sugar separation, saccharates, simulated moving bed, purification, adsorbent, membrane separation processes

Introduction

The term "molasses" is applied for the viscous dark brown syrup obtained in the final crystallization stage of the sugar production process. Molasses represents a by-product from which further sucrose (sugar) separation by direct crystallization is no longer economically justified (McGinnis, 1951). In order to prevent further non-sugars accumulation, molasses is withdrawn from the process (Asadi, 2007). In addition to sugar, the main component (> 48% w/w) (Asadi, 2007), molasses also contains 28 – 33% (w/w) of non-sugar compounds with organic and inorganic origin (Šárka et al., 2013) as well as residual water (~ 20%). The profile of the non-sugar compounds present in molasses differs depending on the sugar beet quality and individual operations applied in the sugar production stages. The non-sugar compounds, particularly colorants, are usually formed during juice evaporation and crystallization (melanoidins, alkaline degradation products of hexoses and caramels) (Coca et al., 2004). Nevertheless, other non-sugars like phenolic compounds, betaine, amino acids, organic acids, are naturally present in sugar beet (Šárka et al., 2013) and during processing are concentrated in molasses. Mentioned non-sugar compounds represent high-added value components with potential application in different industries due to the health protective activity (Chen et al., 2017; Day & Kempson, 2016; Valli et al., 2012). In this regard, separation of sugar and non-sugars present in molasses is a challenge towards complete molasses exploitation and separated compounds valorization.

The quantity and composition of the produced molasses provide an insight to the operations managing during sugar beet processing alongside climate and growing conditions on the beet field (Olbrich, 1963). An average production of molasses is usually 4 – 5% on beet and the corresponding yield is dependent from various factors and differs among campaigns. In Serbia, the overall amount of processed sugar beet during the 2016 campaign was around 4 000 000 t while during 2017 total amount was slightly higher than 3 000 000 t (personal communications). The overall produced sugar beet molasses in the corresponding years was 177 673 t and 137 755 t, respectively and in average close to the value reported by Food and Agriculture Organization of the United Nations (FAO, 2017) in the period 2010–2014 (Figure 1). The quantity of the sugar recovered by the chromatographic process in 2016 campaign was approximately 19 000 t (personal communication). The comparison of produced sugar beet molasses in the leading sugar producing countries in Europe and Serbia for period 2010–2014 is presented in Figure 1.

The loss of sugar in molasses is the largest individual loss of sugar in the production process since 2.2 – 2.6% (w/w) of the sugar beet sugar's content is lagging behind in the molasses (Olbrich, 1963; Schiweck, Clarke & Pollach, 2017). Every factory aims to accomplish equilibrium between sugar yield and commercially acceptable molasses. Produced molasses composition is monitored and regulated by molasses purity quotient.

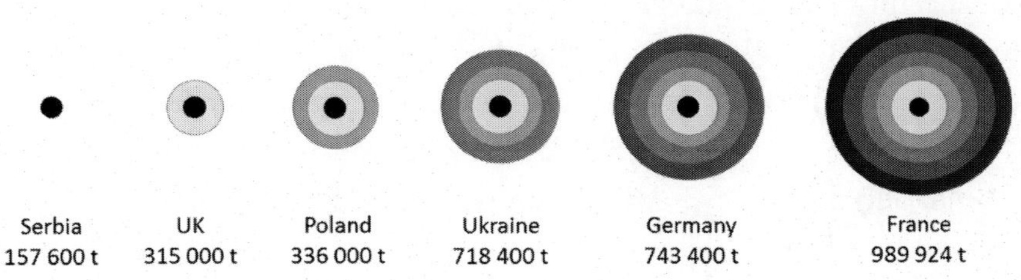

Figure 1. Quantities of the produced sugar beet molasses in European countries in the period 2010–2014 (FAO, 2017) (http://www.fao.org/faostat/en/#data/QD/ visualize).

Typical purity quotient of commonly produced molasses by appropriate processing operations and equipment should not exceed value of 60 (Asadi, 2007; Maudru, 1951). The ratio of sugar to non-sugar compounds (melassigenic coefficient) in molasses with the corresponding purity quotient is usually 1.5. Namely, 1 kg of non-sugars present in molasses is accompanied by 1.5 kg of sugar and hinders its separation (Asadi, 2007). The non-sugars presence causes increase in sucrose solubility and increase in viscosity of the sugar solutions. Consequently, difficulties regarding sucrose crystallization from the corresponding solution are introduced (Kukić, 1995). Water is recognized as the compound with the highest melassigenic affinity followed by potassium and sodium cations and hydroxyl anion (van der Poel, 1998).

Improvement of the existing sugar production process alongside with the development of treatments and methods towards increase in sugar yield has always been in the research focus. The corresponding endeavor has led to the development of processes regarding molasses sugar recovery.

Molasses Desugarization Processes

Previously established processes regarding molasses sugar recovery could be classified in two major groups according to the principle of sugar isolation (Kukić, 1995). The first group comprises of processes where sugar separation is possible by chemical means. These processes are based on sucrose ability to form poorly soluble compounds (saccharates) in reactions with oxides and/or hydroxides of the alkaline-earth elements under proper operating conditions. Besides calcium oxide which is extensively applied and still in use on an industrial scale in this respect (Steffen process), strontium and barium hydroxides were also applied (Strontian and baryte process, respectively) (Hartmann, 1951; Schiweck, Clarke & Pollach, 2017). The second group includes methods based on non-sugars separation and obtainment of sugar solutions with higher purity subsequently processed by conventional operations (evaporation, crystallization). Chromatographic processes alongside osmosis and electrodialysis pertain to this group, but only chromatographic processes were developed enough to be applied in industry (Kukić, 1995). Considering industrial practice, the Steffen process and chromatographic processes are described in the following sections.

Molasses Desugarization by Steffen Process

The first applied molasses desugarization process on an industrial scale was the Steffen process (calcium saccharate process) patented in 1883 by inventor Carl Steffen (Asadi, 2007). Steffen process was primarily introduced and subsequently extensively applied in the United States due to the higher price of crystalline sugar compared to sugar beet molasses (Hartmann, 1951). At present, molasses desugarization by Steffen process is conducted in the United States, the Commonwealth of Independent States (Azerbaijan, Belarus, Georgia, Kazakhstan, Kyrgyzstan, Moldova, Russia, Tajikistan, Turkmenistan, Ukraine, and Uzbekistan) and Middle East countries (Turkey and Iran) (Schiweck, Clarke & Pollach, 2017). Nevertheless, it is gradually suppressed by the development and introduction of the chromatographic processes.

The main principle of the corresponding process is sucrose precipitation in the form of practically insoluble tricalcium saccharate obtained upon calcium oxide (quicklime) addition and proper temperature application. The exact mechanism of the tricalcium saccharate formation is not completely revealed but several theories have been proposed (Hartmann, 1998; Kukić, 1995). The possibility of sugar recovery by applying the mentioned process is estimated at 80 – 85% of the molasses sugar. Nevertheless, if the obtained saccharate milk is returned to the juice purification stage, the actual sugar recovery decreases to 60% due to the increased amount of non-sugar compounds (15%). Considering the given melassigenic coefficient (1.5), the increased amount of non-sugars in juice will carry 25% of sugar back to molasses (Asadi, 2007). Similarly, by application of strontium and barium hydroxides, 90% and 80% of the sugar present in molasses can be recovered (Kukić, 1995).

Schematic presentation of the Steffen process is given in Figure 2. In order to separate sugar from the non-sugar compounds, starting molasses (~ 80% DS) requires dilution. The molasses dilution extent is in the range of 6% dry substance (DS) (Hartmann, 1951) to 15% DS (Schiweck, Clarke & Pollach, 2017) but usually, molasses is diluted to 10 – 12% DS (Perschak, 1998; Kukić, 1995). Under continuous stirring, fine powdered CaO (< 100 μm) is gradually added to the diluted molasses inducing saccharates formation. At the point where 0.7 – 0.9% of free CaO is present in the system, CaO addition is interrupted. By low temperature maintaining (10 – 14°C) during reaction, precipitation (referred as cold precipitation) of the formed tricalcium saccharate is enabled (Schiweck, Clarke & Pollach, 2017). The amount of used CaO is usually 100 – 130% on molasses sugar (Perschak, 1998). Separation of the saccharate precipitate is performed by filtration. In this regard, commonly used are rotary vacuum filters, filter presses or belt filters. Rinsing of the accumulated cold saccharate cake is conducted alongside filtration (Hartmann, 1998; Schiweck, Clarke & Pollach, 2017). The majority of sugar present in molasses is retained in the cold saccharate cake (~ 90%) while small amount passes to the filtrate (~ 10%). After filtrate heating to 90°C the remaining sugar (6.5% of molasses sugar) is precipitated in saccharate form (Hartmann, 1998) and afterwards separated by settling and filtration. The obtained hot saccharate cake is merged with the cold saccharate cake and mixed with sweetwater in order to produce saccharate milk (Perschak, 1998). As regards to filtrate (referred as Steffens waste), it is used for the production of betaine, glutamic acid and raffinose or evaporated and applied as a vinasse in cattle feed (Schiweck, Clarke & Pollach, 2017).

Depending on the factory's preference (available equipment, campaign duration), the obtained saccharate milk is processed in two ways (Perschak, 1998; Schiweck, Clarke & Pollach, 2017):

- returned as a part of the lime milk in juice purification
- subjected to heating (85°C) and carbonation (separately or alongside diffusion juice) after which formed calcium carbonate is separated

by filtration and obtained sugar solution subsequently evaporated and crystallized.

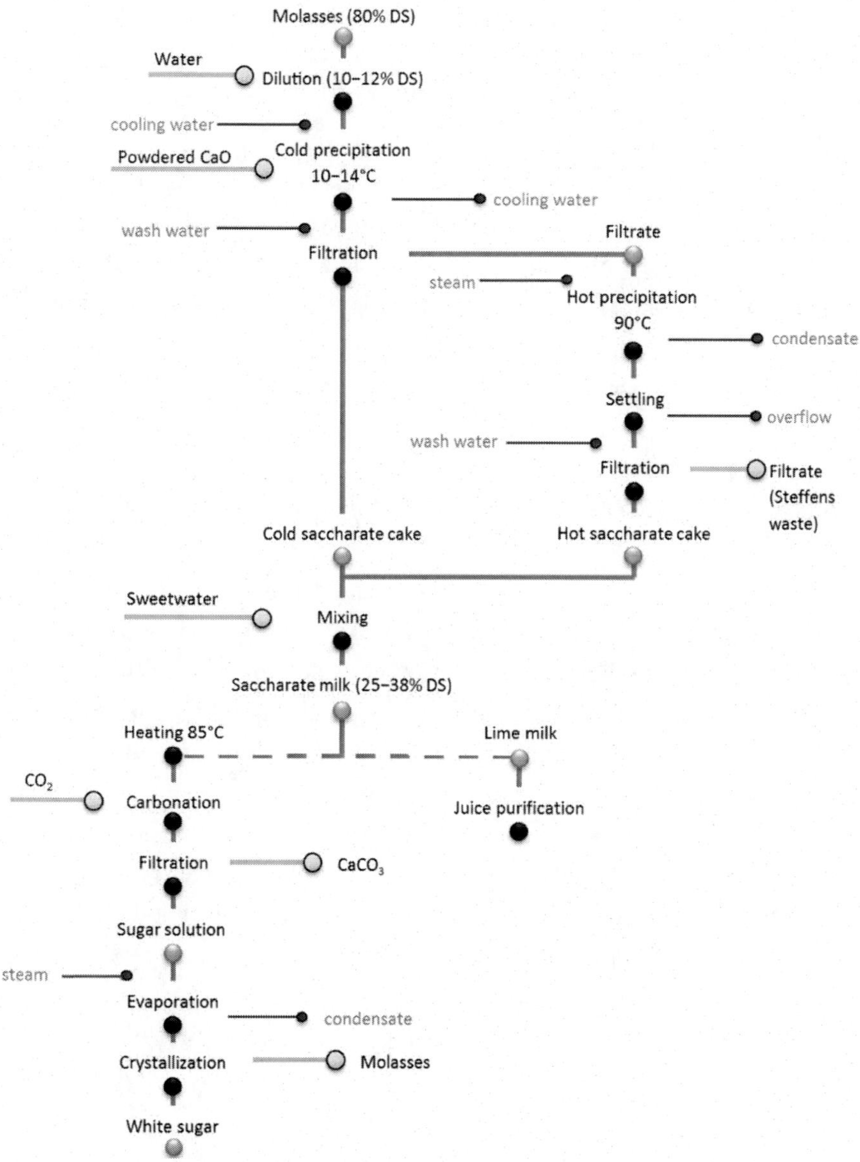

Figure 2. Flow diagram of molasses desugarization by Steffen process.

Sugar recovery by applying the Steffen process is limited to molasses with maximal invert sugar content of 0.4%. High invert sugar content in cane molasses hinders the application of Steffen process (Perschak, 1998). Furthermore, Steffen process enables separation of only one component (sucrose) which is considered as disadvantage. High water consumption and large wastewater amount generation (800 – 1000% on processed molasses) are the main process drawbacks towards growing demand for sustainable environmental protection. Nevertheless, molasses processing regardless of its hardness content as well as low cost and availability of CaO represent positive aspects of the Steffen process (Asadi, 2007; Hartmann, 1998).

Strontian and baryte processes are based on the same principle but were suppressed by Steffen process because of the higher costs and toxicity of the applied hydroxides (Hartmann, 1998).

Molasses Desugarization by Chromatographic Processes

The efforts made to address production and disposal of large wastewater amounts obtained in seminal molasses desugarization processes led to the development and implementation of new technologies. In this respect importance of the molasses desugarization by chromatographic processes (MDC process) has grown rapidly. Ion-exclusion chromatography invented in the 1950s in the United States attracted the sugar technologists' attention. Hongisto and Schoenrock were the pioneers in the investigation of ion-exclusion chromatography application for molasses sugar recovery and subsequently MDC process development. In 1974 the first commercial MDC facility was built in Kantvik (Finland), and in the late 1980s, MDC process has been accepted by beet sugar companies in the United States (Asadi, 2007). Furthermore, in 1997 approximately 80% of molasses produced by US beet sugar industry was processed by MDC process (Schiweck, Clarke & Pollach, 2017).

The main principle of resin-based ion-exclusion chromatography and thus MDC process relies on the ionic compounds exclusion (rejection) and nonionic compounds inclusion (absorption) when treated solution passes through the column filled with a particular resin. In the case of molasses, non-sugars as ionic compounds are excluded and run off first (raffinate) while sucrose as a nonionic compound is retained into the resin bed and later rinsed with water (extract) (Asadi, 2007). The MDC process enables white sugar recovery up to 80%, while approximately 10% of sugar is retained in non-sugar fraction and remaining 10% ends up in secondary molasses (Schiweck, Clarke & Pollach, 2017; Perschak, 1998).

Advantages of MDC process compared to the Steffen process are the possibility for multiple component separation (non-sugar, sucrose and betaine fractions), high process efficiency and wastewater-free processing (Asadi, 2007). However, MDC process application in molasses desugarization requires appropriate molasses pretreatment in order to prevent resin destruction and sugar loss (Perschak, 1998).

Schematic presentation of the MDC processes is given in Figure 3. Firstly, molasses is subjected to pretreatment which comprises of following operations: dilution, heating, softening, filtration, and degassing. Molasses diluted to 60% DS using condensate is heated to 80°C and subsequently softened. Diluted molasses softening is often necessary, considering that bivalent ion content should not exceed 1.5 mmol/100 g molasses DS. Softening is usually conducted either by sodium carbonate addition or ion exchange treatment. After softening, suspended solids are separated by filtration in order to prevent resin fouling. Resin destruction by oxidative de-cross-linking and channeling in the resin bed induced by gas bubbles is inhibited by diluted molasses degassing (Perschak, 1998). Diluted molasses obtained after described pre-treatment enters the chromatographic columns where the separation of non-sugars is taking place. At this point, four different continuous large scale commercial separation processes are available:

- simulated moving bed (SMB)
- sequential simulated moving bed (sequential SMB)
- coupled loop chromatography (CLC) or ARI system developed by Amalgamated Research Inc., USA

- FAST system developed by Finnsugar and Applexion Separation Technology.

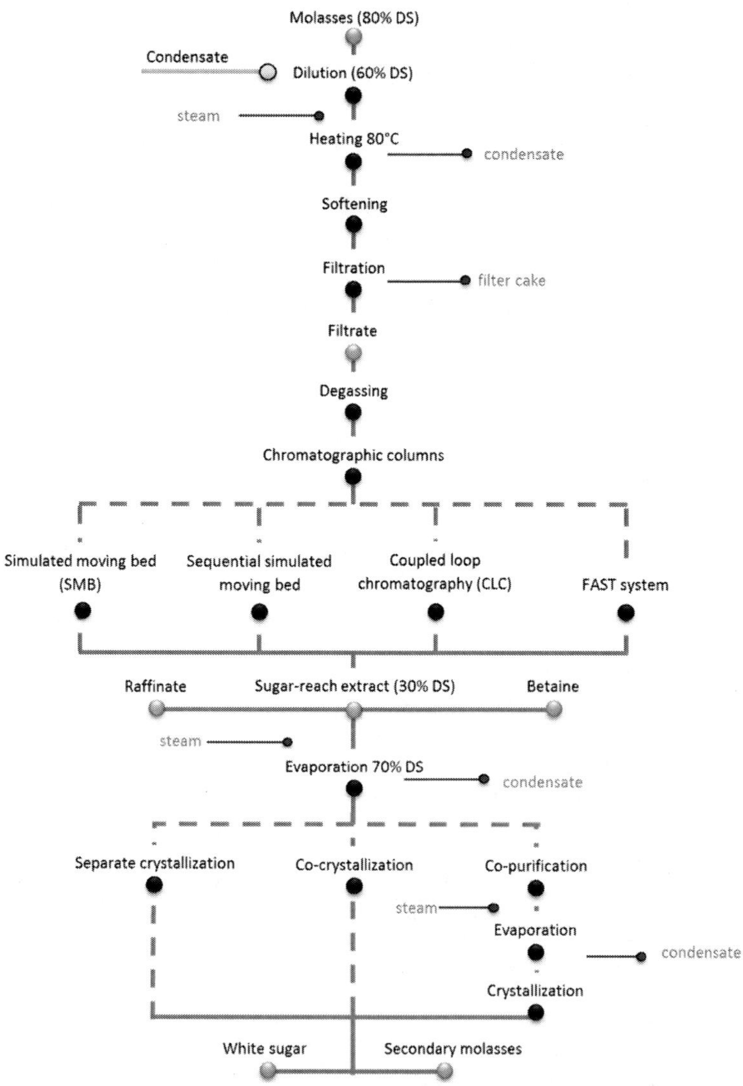

Figure 3. Flow diagram of molasses desugarization by chromatographic processes.

The basis of all mentioned large scale commercial processes is simulated moving bed chromatographic technique introduced in the late 1950s and applied in many areas (fine chemistry, biotechnology, pharmaceuticals, food and sugar processing, petrochemical industry) (Hassan et al., 2016). The corresponding technique is named SMB considering that the resin movement is simulated by the periodical changes in the position of inputs (molasses and water) and outputs (extract and raffinate) flow from one column to another.

A typical continuous SMB system consists of eight columns in which the position of each input and output simultaneously changes to the next column at a specific time interval. Inputs and outputs are continuously introduced or withdrawn from the columns (Asadi, 2007). The standard SMB chromatography allows only separation of two components (sugar and non-sugars) (Hassan et al., 2016). To overcome this limitation and ensure multiple components separation further enhancement in the performance of the standard SMB system led to the development of sequential SMB, coupled loop chromatography (ARI system) and FAST system.

Sequential SMB chromatography developed by Heikkila et al. (1989) is a technique in which the flows of inputs (molasses and water) and outputs (extract and raffinate) are interrupted sequentially during the operational cycle at specific intervals. A typical sequential SMB system consists of three columns with enabled liquid flow around the column system during one circulation step and formation of a closed loop. Inputs (molasses and water) are introduced at different points in the system while the outputs (extract and raffinate) are discontinuously collected at fixed points. Sequential SMB technique allows separation of three components (sugar, non-sugars and betaine) with high efficiency (Paananen, 1997).

Coupled loop chromatography (CLC) as the further enhancement of ion-exclusion chromatography was introduced by Amalgamated Research Inc. (ARI system) in 1997 (Asadi, 2007). The CLC system consists of several internally independent but externally co-dependent chromatographic sections (recycle loops). The number of loops in the network is always more than one and is determined by the number of desired components to be separated (Perschak, 1998). For sugar recovery CLC system with two loops is usually used. Commonly, loop 1 includes four cells and conditions that constrain sucrose to behave as an excluded component and move alongside with the non-sugars producing the upgrade fraction and betaine fraction as outputs. Afterwards, the concentrated upgrade fraction (sucrose and non-sugars) enters the loop 2 (comprised of eight cells) resulting in the obtainment of sugar-rich extract ($\geq$ 92% purity) and raffinate (12% purity) (Asadi, 2007).

The FAST system represents an enhanced version of the sequential SMB system introduced in the 1990s in the United States. Similarly to sequential SMB system, the basis of FAST system is a three column sequential process which differs in fixed molasses input (into the feeding column) at specific time intervals (Paananen, 1997).

Sugar-rich extract, raffinate rich in non-sugars and betaine fraction are the main outputs obtained after application of the aforementioned large scale commercial chromatographic separation processes. Raffinate is collected first considering that non-sugars are mainly ionic compounds rejected from the resin and thus forced to move quickly through the column. The obtained raffinate (about 6% DS and 12 − 20% purity) is further evaporated to 60% DS and used as cattle feed. Sugar together with other nonionic compounds retained in the resin bed is removed from the columns by rinsing with water. The obtained sugar-rich extract (about 30% DS and 90 − 93% purity) is further evaporated to 70% DS (Perschak, 1998). Processing of concentrated sugar-rich extract into sugar and secondary molasses (extract molasses) could be conducted in three different ways (Asadi, 2007):

- by separate crystallization
- by co-crystallization
- by co-purification

Separate crystallization refers to concentrated sugar-rich extract processing outside campaign by conventional crystallization. During the beet campaign, concentrated sugar-rich extract could be processed either by co-crystallization or co-purification. In the co-crystallization, concentrated sugar-rich extract is mixed with standard liquor obtained in beet processing and further processed by conventional crystallization. In the co-purification process, concentrated sugar-rich extract enters the prelimer to be limed with diffusion juice and further processed by conventional processing. Secondary molasses amount generated after concentrated sugar-rich extract processing is 10 − 12% calculated on desugarized molasses (Asadi, 2007).

MOLASSES PURIFICATION TREATMENTS

Conventional juice purification is based on the non-sugar compounds precipitation and destabilization by lime and carbon dioxide (carbonation gas) addition. Through many years of industrial practice, different juice purification systems were developed (the most frequently used BMA, DDS, Dorr, Novi Sad) (Asadi, 2007; Madsen, 1998; Maravić et al., 2015). In addition, raw juice can be decolorized by using ion-exchange resin, large surface area adsorbents (bone char, activated carbon) or sulfur dioxide (sulfitation) (Asadi, 2007). Over the past years, extensive research regarding membrane separation processes application (microfiltration, ultrafiltration, nanofiltration, reverse osmosis and electrodialysis) in juice purification was also conducted (Bhattacharya et al., 2001; Hinkova et al., 2002; Jegatheesan et al., 2012; Lutin, Bailly & Barb, 2002; Madsen, 1998; Tragardh and Gekas, 1988). After purification, the non-sugars content in raw juice is reduced by 20 to 30% enabling the production of white sugar with specific purity (Asadi, 2007). However, the remaining non-sugars and colorants formed during further juice processing are accumulated in molasses. Non-sugars presence in molasses often causes difficulties in sugar recovery as well as in its further application as the raw material for a wide range of industries (food, feed and fermentation industries). In this regard, non-sugars removal is preferable and hence imposes introduction of molasses purification treatment. Detailed investigations on possible molasses purification treatments are scarce. It is assumed that treatments applied in juice purification could be used in molasses purification as well. Research conducted on the adsorbents application (particularly bentonite) in molasses purification is described in the following section. Furthermore, molasses purification by membrane separation processes application alone or coupled with adsorbent treatment is presented.

Molasses Purification by Bentonite Addition

Potential of adsorbents application in molasses purification has not been intensively explored as in juice purification. Adsorbents used for juice purification, in terms of decolorization, colloids, floc and foaming agents' removal, are bone char, activated carbon (Perschak, 1998) and bentonite (Erdogan et al., 1996; Jahed et al., 2014; Laksameethanasana et al., 2012). Bone char and activated carbon are introduced on an industrial scale while the bentonite application is still in the domain of laboratory research. At present, juice purification with bone char is not extensively employed except in cane sugar refining. Activated carbon is used in powdered or granular form, as a body feed or as a precoat (Perschak, 1998). With bentonite application in raw sugarcane (Laksameethanasana et al., 2012) and sugar beet juice (Jahed et al., 2014) purification treatment, great improvements in color and turbidity reduction were achieved.

Considering promising results accomplished in juice purification, adsorbents application in molasses purification treatment represents a field for further research. At present, bentonite adsorption potential in molasses purification was the subject of two studies (Djordjević et al., 2017; Djordjević et al., 2018a). The large surface area alongside high cation exchange capacity as bentonite natural characteristics enables outstanding adsorption properties towards various types of cations and organic pollutants (Shaikh et al., 2017). Bentonite represents colloidal aluminum-silicate clay primarily comprised of mineral montmorillonite (at least 70%) with a dominant exchangeable ion which determines bentonite type (Savic et al., 2014). Sodium (Na), calcium (Ca) and combined Na-Ca bentonites are commonly applied bentonite types in the industry. In the corresponding studies, four different bentonites were employed: Claris p30 and Claris p50 as combined type (Djordjević et al., 2017), Na Bent and Ca Bent (Djordjević et al., 2018a). Bentonite was added to molasses as suspension in order to reach concentrations of 9, 15, 21 g/L. The influence of treatment conditions such as pH (3, 5, 7) and molasses DS (40, 50, 60° Brix) was also examined. Experiments were conducted according to the Box-Behnken experimental design (Table 1).

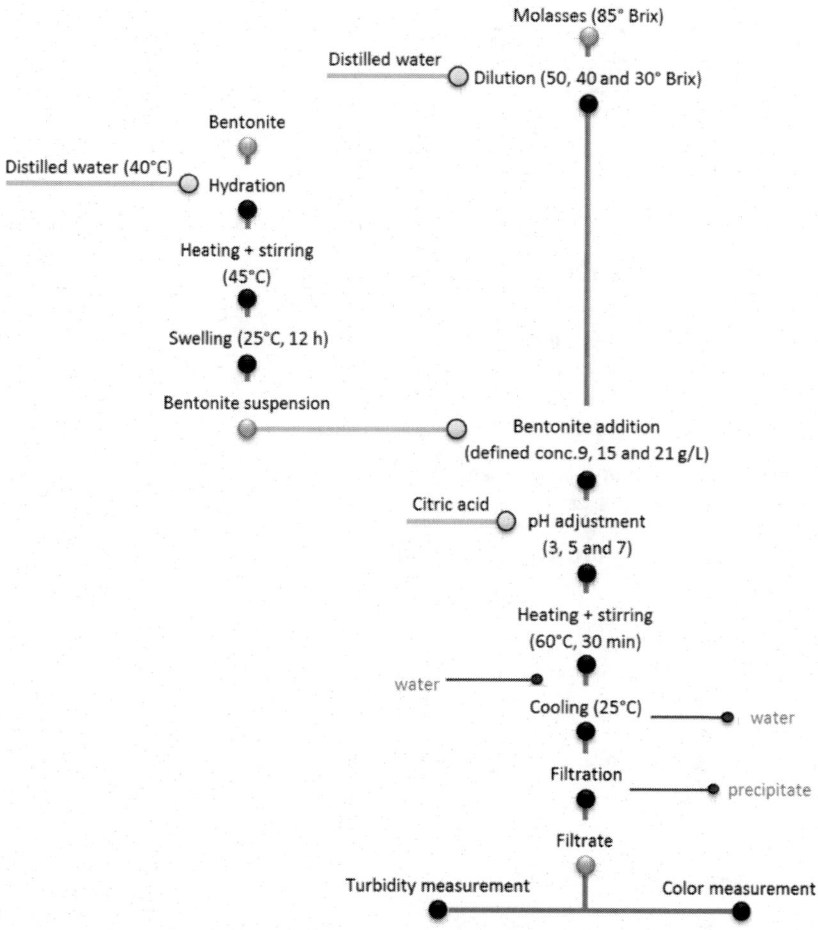

Figure 4. Flow diagram of the molasses purification by bentonite addition.

Table 1. Summary of the results obtained regarding molasses color reduction in the molasses purification treatment upon different bentonites application according to the employed Box-Behnken design

Run	Independent variables			Response-Color reduction* [%]			
	pH	Molasses DS [° Brix]	Bentonite conc. [g/L]	Claris p30	Claris p50	Ca Bent	Na Bent
1	3.0	40	9	21.75	40.57	35.39	43.74
2	7.0	40	9	7.92	26.53	14.18	4.20
3	3.0	40	21	17.51	42.86	46.10	48.43
4	7.0	40	21	6.66	24.92	19.91	18.91
5	3.0	30	15	22.34	48.93	28.00	35.01
6	7.0	30	15	1.51	31.62	0.11	15.23
7	3.0	50	15	19.34	38.29	23.86	36.53
8	7.0	50	15	0.22	14.06	1.94	20.36
9	5.0	30	9	0.12	24.31	31.36	8.11
10	5.0	30	21	11.54	20.60	34.36	23.16
11	5.0	50	9	10.18	25.68	23.84	8.48
12	5.0	50	21	18.14	38.17	17.65	6.23
13	5.0	40	15	5.14	33.02	11.80	32.45
14	5.0	40	15	2.38	19.25	12.53	43.74
15	5.0	40	15	3.22	27.23	16.13	44.74

* Values are the mean of two replicates

Source: Djordjević et al., 2017; Djordjević et al., 2018a

The molasses purification treatment efficiency was assessed by molasses color and turbidity measurements. Schematic presentation of the molasses purification by bentonite addition is given in Figure 4. Performed bentonite treatment consisted of bentonite suspension preparation, molasses dilution, bentonite suspension addition, pH adjustment, heating with stirring, subsequent cooling and filtration. Bentonite suspensions were prepared by hydration of bentonite powder with required water amount (40°C – 50°C) followed by intense stirring until uniform suspension formation. Subsequently, swelling of the obtained suspension is enabled during 12 h at 25°C. The adequate amount of bentonite suspension was added to diluted molasses (200 ml, 30, 40, 50° Brix) in order to achieve previously defined concentrations (9, 15, 21 g/L). Defined pH (3, 5, 7) was adjusted by citric acid addition. Obtained blends were further subjected to heating with constant stirring in a water bath at 60°C during 30 min. After cooling, blends were filtered through filter paper and the obtained filtrate was used for color and turbidity assessment. The obtained results regarding molasses color and turbidity reduction achieved by described treatment conducted with different bentonites are summarized in Tables 1 and 2.

Investigation of Claris p30 and Claris p50 performance on molasses color and turbidity reduction reveal greater Claris p50 reduction efficiency. Color and turbidity reduction extent was greatly dependent on pH and bentonite concentrations. The most efficient color reduction was accomplished in the acidic conditions (pH 3, pH 5) with increasing bentonite concentration. Neutral pH alongside medium and high bentonite concentration were established as suitable for maximal turbidity reduction (Djordjević et al., 2017).

Table 2. Summary of the results obtained regarding molasses turbidity reduction in the molasses purification treatment upon different bentonites application according to the employed Box-Behnken design

Run	Independent variables			Response-Turbidity reduction* [%]			
	pH	Molasses DS [° Brix]	Bentonite conc. [g/L]	Claris p30	Claris p50	Ca Bent	Na Bent
1	3.0	40	9	0	0.63	68.82	84.87
2	7.0	40	9	66.97	48.90	85.37	92.76
3	3.0	40	21	0	55.04	89.55	98.58
4	7.0	40	21	90.20	92.65	89.47	99.38
5	3.0	30	15	30.87	97.45	64.63	90.94
6	7.0	30	15	67.65	99.08	85.00	93.08
7	3.0	50	15	0	0.50	87.16	93.96
8	7.0	50	15	80.31	52.77	85.62	90.04
9	5.0	30	9	0	97.08	63.83	87.81
10	5.0	30	21	57.32	98.37	77.98	97.33
11	5.0	50	9	25.46	1.12	94.72	90.45
12	5.0	50	21	54.59	63.27	94.15	97.15
13	5.0	40	15	47.75	54.73	91.46	96.74
14	5.0	40	15	63.35	73.34	87.61	97.55
15	5.0	40	15	45.88	64.40	90.06	98.50

* Values are the mean of two replicates

Source: Djordjević et al., 2017; Djordjević et al., 2018a

In the study which involved the use of Ca Bent and Na Bent more efficient color and turbidity reduction was achieved upon Na Bent addition. Acidic conditions (pH 3) were established as more appropriate in terms of color reduction while greater turbidity reduction was accomplished in the neutral pH. However, applied acidic conditions lead to molasses sucrose content reduction (due to sucrose inversion). Higher efficiency in color and turbidity reduction was also observed with increasing bentonite concentration while the influence of molasses DS was not strongly expressed. The optimal treatment conditions which allow maximal color and turbidity reduction with minimal sucrose loss were: pH 5.11 – 5.17, molasses DS 38 – 40° Brix and bentonite concentration 16 – 21 g/L (Djordjević et al., 2018a).

Molasses Purification by Membrane Separation Processes

The simple principle of compounds rejection by the membrane, defined by membrane pore size and compound molecular size has enabled widespread application of membrane separation processes in various industries. Depending on the desired compound, the resulting streams, retentate (concentrate) and permeate, are retained or discarded. According to the membrane pore size, following membrane separation processes are distinguished: microfiltration (MF), ultrafiltration (UF), nanofiltration (NF), reverse osmosis (RO) and electrodialysis (ED) (Galanakis et al., 2016).

Research involving membrane separation processes application in the sugar industry started in the early 1970s. MF and UF were among the most commonly applied membrane separation processes in terms of non-sugar compounds removal from the sugar industry intermediate products (syrups and juices) as evidenced by numerous studies (Hakimzadeh et al., 2006; Hamachi et al., 2003; Jegatheesan et al., 2009; Li et al., 2016; Luo et al., 2016; Šereš et al., 2008; Shahidi & Razavi, 2006). In this regard, ceramic membranes, as well as polymeric membranes, were employed. In order to enhance permeate flux during sugar beet syrup purification, static mixer was also introduced (Šereš et al., 2010).

However, only few studies investigated the possibility of molasses purification by using MF and UF (Guo et al., 2018; Jones et al., 2012). Nevertheless, attempts have been made to apply the corresponding membrane separation processes coupled with adsorbents treatment for molasses purification (Bernal et al., 2016; Djordjević et al., 2018b).

Jones et al. (2012) subjected sugar beet molasses to cross-flow MF with the aim to determine and optimize the cleaning process for the fouled membranes. The polymeric membranes employed were made of polysulphone with 0.5, 0.9 and 1.5 μm pore size. Sugar beet molasses was diluted to 45° Brix and filtered under the following conditions: temperature 60°C, transmembrane pressure 300 kPa, feed flow rate of 1.89 m/s during 90 min. As reported by the authors, polysulphone membrane with 1.5 μm pore size was found to be suitable for molasses treatment with fouling fluxes of 56(±3) L/m^2h and acceptable compounds removal. Unfortunately, results regarding potential improvement in molasses color and turbidity were not reported.

An interesting study regarding sugar cane molasses decolorization using cross-flow UF was conducted by Guo et al. (2018). Polymeric membranes (polyamide and polyether sulphone) with different molecular weight cut-offs (2kD-PA, 3kD-PES, and 6kD-PES) in spiral-wound membrane module were employed and their effect in terms of sugar cane molasses color and turbidity reduction have been evaluated. The influence of molasses dilution extent (1:2, 1:3 and 1:4 g/g) and examined pH (5, 6 and 7) on the molasses quality parameters was also studied. The applied conditions were constant transmembrane pressure of 1000 kPa, feed flow rate of 5 L/min and temperature of 50 and 60°C. In order to mitigate reversible and irreversible membrane fouling, molasses pretreatment and alkaline treatment on UF polymeric membrane were investigated. Three different molasses pretreatments were applied, namely UF with a ceramic membrane, centrifugation, and chitosan flocculation. After decolorization, the obtained UF permeate was subsequently used for sugar recovery by NF. According to the authors, 2kD-PA membrane was adequate for molasses decolorization with achieved color and turbidity reduction of 77 – 89% and 100%, respectively. As regards to molasses pretreatment, centrifugation was marked as most suitable considering process costs. Membrane alkaline treatment was estimated as appropriate for improvements in anti-fouling and permeate flux. The optimal conditions for molasses decolorization highlighted by the authors were neutral pH and a temperature of 60°C. Reported sucrose recovery in the corresponding UF permeate at 60°C was 96%. The authors suggest that the presented method provides guidance towards industrial application.

The coupled effect of ultrafiltration and activated carbon on sugar beet molasses decolorization with subsequent activated carbon regeneration was investigated by Bernal et al. (2016). The effects of treatment conditions such as transmembrane pressure (50 and 100 kPa), feed flow rate (1.86 and 4.24 L/h), feed pH (1, 3 and 7.1) and initial activate carbon concentration (1.5 – 5 g/L) on molasses color reduction, permeate flux and membrane fouling were assessed. Diluted molasses (50 g/L) containing added powdered activated carbon was subjected to UF at 25°C through tubular ceramic membrane (Tami Industries, cut-off from 10 to 300 kDa). According to the authors, molasses color reduction was primarily associated with colorants adsorption onto powdered activated carbon. Molasses color reduction of 96.5% was accomplished at optimal conditions: pH 3, activated carbon concentration 5 g/L, transmembrane pressure 100 kPa and a feed flow rate of 4.24 L/h.

Recently, Djordjević et al. (2018b) studied the effects of combined Na Bent pretreatment and cross-flow MF on sugar beet molasses color and turbidity reduction. The presented study also involved the use of a static mixer (stainless steel twisted tape) aiming to enhance permeate flux. Experiments were conducted in static mixer absence or presence at constant transmembrane pressure of 200 kPa according to the 2^2 factorial experimental design with independent variables feed flow rate (100 and 300 L/h) and molasses DS (40 and 50° Brix). Molasses pretreatment consisted of Na Bent suspension addition in order to achieve defined concentration of 7 g/L, pH adjustment (pH 5), heating with stirring (50°C for 30 min.), and subsequent filtration through cellulose filter cloth. The obtained filtrate was subjected to MF through single-channel ceramic membrane (200 nm pore size, GEA, Westfalia, Germany) in the static mixer absence and presence at

50°C. Compared to static mixer presence, in static mixer absence, higher molasses color (20 − 60%) and turbidity reduction (83.4 − 99.2%) was achieved. Nevertheless, static mixer application increased the final and steady-state permeate fluxes but reflected unfavorable on molasses color and turbidity reduction.

CONCLUSION

Molasses desugarization and purification by employing different processes and treatments has been the subject of research for over a century. During this period numerous processes were developed and applied on an industrial scale. Furthermore, demands for greater sucrose recovery and rising environmental concerns shifted the scientific research focus towards implementation of more sophisticated technologies. In this regard chromatographic and membrane separation processes were introduced. At present, sugar recovery on an industrial scale is accomplished by various chromatographic processes. Compared to chromatography, the scarce results regarding membrane separation processes and adsorbents application in molasses treatment limit their industrial implementation. However, it is important to emphasize that further enhancement of the corresponding technologies is still under intensive scientific research.

REFERENCES

Asadi, M. (2007). *Beet-Sugar Handbook*. Hoboken, NJ: John Wiley & Sons.

Bernal, M., Ruiz, M. O., Geanta, R. M., Benito, J. M. & Escudero, I. (2016). Colour removal from beet molasses by ultrafiltration with activated charcoal. *Chemical Engineering Journal*, *283*, 313–322. http://doi.org/10.1016/j.cej.2015.07.047.

Bhattacharya, P. K., Agarwal, S., De, S. & Rama Gopal, U. V. S. (2001). Ultrafiltration of sugar cane juice for recovery of sugar: analysis of flux and retention. *Separation and Purification Technology, 21*, 247–259.

Chen, M., Zhao, Z., Meng, H. & Yu, S. (2017). The antibiotic activity and mechanisms of sugar beet (*Beta vulgaris*) molasses polyphenols against selected food-borne pathogens. *LWT-Food Science and Technology*, *82*, 354–360. http://dx.doi.org/10.1016/j.lwt.2017.04.063.

Coca, M., Garcia, M. T., Gonzalez, G., Pena, M. & Garcia, J. A. (2004). Study of coloured components formed in sugar beet processing. *Food Chemistry*, *86*, 421–433. https://doi.org/10.1016/j.foodchem.2003.09.017.

Day, C. R. & Kempson, S. A. (2016). Betaine chemistry, roles, and potential use in liver disease. *Biochimica et Biophysica Acta*, *1860*, 1098–1106.

Djordjević, M., Šereš, Z., Djordjević, M., Šoronja-Simović, D., Maravić, N. & Došenović, T. (2017). Bentonite as environmental friendly sorption material for sugar beet molasses purification. In *Proceedings of the 23rd International Symposium on Analytical and Environmental Problems*, Szeged, Hungary, 9–10 October 2017. https://doi.org/10.5281/ZENODO.1065312.

Djordjević, M., Šereš, Z., Došenović, T., Šoronja-Simović, D., Maravić, N., Kukić, D., Nikolić, I. & Djordjević, M. (2018a). Sugar beet molasses purification by bentonite addition: Analysis of quality enhancement and treatment conditions. *LWT-Food Science and Technology*, *93*, 142–149. https://doi.org/10.1016/j.lwt.2018.03.030.

Djordjević, M., Šereš, Z., Došenović, T., Šoronja-Simović, D., Maravić, N., Šaranović, Ž, Šereš, L., Šćiban, M. & Djordjević, M. (2018b). Modelling of cross-flow microfiltration coupled with bentonite treatment in sugar beet molasses purification. *The Canadian Journal of Chemical Engineering*. http://dx.doi.org/10.1002/cjce.23289.

Erdogan, B., Demirci, Ş. & Akay, Y. (1996). Treatment of sugar beet juice with bentonite, sepiolite, diatomite and quartamin to remove color and turbidity. *Applied Clay Science*, *11*, 55–67. http://doi.org/10.1016/0169-1317(96)00012-9.

Food and Agriculture Organization of the United Nations-FAO. (2017). Production quantities of Molasses by country. *FAO*. Retrieved from http://www.fao.org/faostat/en/#data/QD/visualize.

Galanakis, C. M., Castro-Muñoz, R., Cassano, A. & Conidi, C. (2016). Recovery of high-added-value compounds from food waste by membrane technology. In A. Figoli, A. Cassano, & A. Basile (Eds.), *Membrane technologies for biorefining*, (pp. 189–215). http://dx.doi.org/10.1016/B978-0-08-100451-7.00008-6.

Guo, S., Luo, J., Wu, Y., Qi, B., Chen, X. & Wan, Y. (2018). Decoloration of sugarcane molasses by tight ultrafiltration: Filtration behavior and fouling control. *Separation and Purification Technology*, *204*, 66–74. https://doi.org/10.1016/j.seppur.2018.04.067.

Hakimzadeh, V., Razavi, S. M. A., Piroozifarda, M. Kh. & Shahidi, M. (2006). The potential of microfiltration and ultrafiltration process in purification of raw sugar beet juice. *Desalination*, *200*, 520–522. http://doi.org/10.1016/j.desal.2006.03.420.

Hamachi, M., Gupta, B. B. & Ben Aim, R. (2003). Ultrafiltration: a means for decolorization of cane sugar solution. *Separation and Purification Technology*, *30*, 229–239.

Hartmann, E. M. (1951). The saccharate processes: The Steffen process. In R. A. McGinnis (Ed.), *Beet-Sugar Technology*, (pp. 438–452). New York, USA: Reinhold Publishing Corporation.

Hassan, N., Henke, S., Hinkova, A., Pour, V. & Bubník, Z. (2016). Molasses separation using improved SMB distributor. *Journal of Food Engineering*, *172*, 19–24. http://dx.doi.org/10.1016/j.jfoodeng.2015.06.008.

Heikkila, H., Hyoky, G. & Kuisma, J. (1989). *Method for the recovery of betaine from molasses*, European Patent 0345511 A2.

Hinkova, A., Bubnik, Z., Kadlec, P. & Pridal, J. (2002). Potentials of separation membranes in the sugar industry. *Separation and Purification Technology*, *26*, 101–110.

Jahed, E., Khodaparast, M. H. H. & Mousavi Khaneghah, A. (2014). Bentonite, temperature and pH effects on purification indexes of raw sugar beet juice to production of inverted liquid sugar. *Applied Clay Science*, *102*, 155–163. http://doi.org/10.1016/j.clay.2014.09.036.

Jegatheesan, V., Phong, D. D., Shu, L. & Ben Aim, R. (2009). Performance of ceramic micro- and ultrafiltration membranes treating limed and partially clarified sugar cane juice. *Journal of Membrane Science*, *327*, 69–77. http://doi.org/10.1016/j.memsci.2008.11.008.

Jegatheesan, V., Shu, L., Keir, G. & Phong, D. D. (2012). Evaluating membrane technology for clarification of sugarcane juice. *Reviews in Environmental Science and Biotechnology*, *11*, 109−124. http://dx.doi.org/10.1007/s11157-012-9271-1.

Jones, S. A., Chewa, Y. M. J., Wilson, D. I. & Bird, M. R. (2012). Fluid dynamic gauging of microfiltration membranes fouled with sugar beet molasses. *Journal of Food Engineering*, *108*, 22−29. http://dx.doi.org/10.1016/j.jfoodeng.2011.08.001.

Kukić, G. (1995). Melasa. U S. Šušić, S. Petrov, G. Kukić, V. Sinobad, P. Perunović, B. Koronsovac, i Đ. Bašić (Ur.), *Osnovi tehnologije šećera II*, (drugo izdanje, pp. 193−265). Beograd, Jugoslavija: Industrija šećera SR Jugoslavije i "Jugošećer". [The molasses. In S.Šušić, S. Petrov, G. Kukić, V. Sinobad, P. Perunović, B. Koronsovac, & Đ. Bašić (Eds.), *Basics of sugar technology II*, (2nd edition, pp. 193−265). Belgrade, Yugoslavia: Yugoslavian sugar industry and "Jugošećer"].

Laksameethanasana, P., Somla, N., Janprem, S. & Phochuen, N. (2012). Clarification of sugarcane juice for syrup production. *Procedia Engineering*, *32*, 141–147. http://doi.org/10.1016/j.proeng.2012.01.1248.

Li, W., Ling, G. Q., Huang, P., Li, K., Lu, H. Q., Hang, F. X., Zhang, Y., Xie, C. F., Lu. D. J., Li, H., Liang, X. Q. & Xiang, J. H. (2016). Performance of ceramic microfiltration membranes for treating carbonated and filtered remelt syrup in sugar refinery. *Journal of Food Engineering*, *170*, 41−49. http://dx.doi.org/10.1016/j.jfoodeng.2015.09.012.

Luo, J., Hang, X., Zhai, W., Qi, B., Song, W., Chen, X. & Wan, Y. (2016). Refining sugarcane juice by an integrated membrane process: Filtration behavior of polymeric membrane at high temperature. *Journal of*

Membrane Science, *509*, 105–115. http://dx.doi.org/10.1016/j.memsci.2016.02.053.

Lutin, F., Bailly, M. & Barb, D. (2002). Process improvements with innovative technologies in the starch and sugar industries. *Desalination*, *148*, 121−124.

Madsen, R. F. (1998). Juice purification: Juice purification systems. In P. W. van der Poel, H. M Schiweck, & T. K. Schwartz (Eds.), *Sugar technology: Beet and Cane sugar manufacture*, (pp. 500–523). Berlin, Germany: Verlag Dr. Bartens.

Maravić, N., Kiss, F., Šereš, L., Bogdanović, B., Bogdanović, B. & Šereš, Z. (2015). Economic analysis and LCA of an advanced industrial-scale raw sugar juice purification procedure. *Food and Bioproducts Processing*, *95*, 19–26. http://dx.doi.org/10.1016/j.fbp.2015.02.004.

Maudru, J. E. (1951). Juice purification: Adsorbing and absorbing agents: Activated carbon. In R. A. McGinnis (Ed.), *Beet-Sugar Technology*, (pp. 289−310). New York, USA: Reinhold Publishing Corporation.

McGinnis, R. A. (1951). Preliminaries to processing. In R. A. McGinnis (Ed.), *Beet-Sugar Technology*, (pp. 123−138). New York, USA: Reinhold Publishing Corporation.

Olbrich, H. (1963). The Molasses. Berlin: Institut für Zuckerindustrie. *Reedition Biotechnologie-Kempe* GmbH 2006. http://www.btkempe.de/Molasses_OLBRICH.pdf, 9.12.2016.

Paananen, H. (1997). Trends in the chromatographic separation of molasses. *Zuckerindustrie*, *122*, 23−33.

Perschak, F. (1998). Molasses desugarizing. In P. W. van der Poel, H. M Schiweck, & T. K. Schwartz (Eds.), *Sugar technology: Beet and Cane sugar manufacture*, (pp. 939–950). Berlin, Germany: Verlag Dr. Bartens.

Savic, I., Stojiljkovic, S., Savic, I. & Gajic, D. (2014). Industrial application of clays and clay minerals. In L. R. Wesley (Ed.), *Clays and clay minerals: Geological origin, mechanical properties and industrial applications*, (pp. 379–402). New York: Nova Science Publishers Inc.

Schiweck, H., Clarke, M. & Pollach, G. (2017). Sugar. In: *Ullmann's Feed and Feed*, (pp. 915–987). Weinheim, Germany: Wiley-VCH Verlag GmbH &Co.

Šárka, E., Bubnik, Z., Hinkova, A., Gebler, J. & Kadlec, P. (2013). Beet molasses – desugarization, composition, properties and application possibilities. *Sugar Industry*, *138*, 105–114.

Šereš, Z., Gyura, J., Djurić, M., Vatai, G. & Eszterle, M. (2008). The application of membrane separation processes as environmental friendly methods in the beet sugar production. In E. Burcu Ozkaraova Gungor (Ed.), *Environmental Technologies*, (pp. 193–208). Rijeka, Croatia: InTech. http://www.intechopen.com/books/environmental_technologies/the_application_of_membrane_separation_processes_as_environmental_friendly_methods_in_the_beet_sugar.

Šereš, Z., Gyura, J., Djurić, M., Vatai, G. & Jokić, A. (2010). Separation of non-sucrose compounds from sugar-beet syrup by ultrafiltration with ceramic membrane containing static mixer. *Desalination*, *250*, 136–143. http://dx.doi.org/10.1016/j.desal.2008.10.034.

Shahidi, M. & Razavi, S. M. A. (2006). Improving thin sugar beet juice quality through ultrafiltration. *Desalination*, *200*, 518–519. http://doi.org/10.1016/j.desal.2006.03.419.

Shaikh, S. M. R., Nasser, M. S., Hussein, I. A. & Benamor, A. (2017). Investigation of the effect of polyelectrolyte structure and type on the electrokinetics and flocculation behavior of bentonite dispersions. *Chemical Engineering Journal*, *311*, 265–276. http://doi.org/10.1016/j.cej.2016.11.098.

Tragardh, G. & Gekas, V. (1988). Membrane Technology in the Sugar Industry. *Desalination*, *69*, 9–17.

Valli, V., Gómez-Caravaca, A., Di Nunzio, M., Danesi, F., Caboni, M. & Bordoni, A. (2012). Sugar cane and sugar beet molasses, antioxidant-rich alternatives to refined sugar. *Journal of Agricultural and Food Chemistry*, *60*, 12508–12515. http://doi.org/ 10.1021/jf304416d.

Van der Poel (1998). Crystallization: Influence of individual nonsugars on molasses formation. In P. W. van der Poel, H. M Schiweck, & T. K. Schwartz (Eds.), *Sugar technology: Beet and Cane sugar manufacture*, (pp. 685–686). Berlin, Germany: Verlag Dr. Bartens.

In: Molasses: Forms, Production and Uses ISBN: 978-1-53614-703-2
Editors: K. Maddison and R. Fuller

Chapter 4

MOLASSES PRODUCTION AND UTILIZATION IN CAMEROON

Fernand Tendonkeng[1,*], Emile Miegoue[1], Bienvenu Fogang Zogang[2] and Etienne Pamo Tedonkeng[1]

[1]Department of Animal Sciences, FASA, University of Dschang, Dschang, Cameroon

[2]ENSAI, Université de Ngaoundéré, Ngaoundéré, Cameroun

ABSTRACT

This chapter reviews the state of knowledge on the production, chemical composition and uses of sugar cane molasses in animal feeding in Cameroon and briefly examine its other uses. In this country, sugar cane molasses, the main sub-product of sugar industries, is mainly produced by the sugar company in Cameroon (SOSUCAM). Formerly used mainly in the fertilization of plantations, it is nowadays a multipurpose by-product. It is used in the production of bioethanol, feed for livestock, pharmaceutical industry and energy production. With a water content of 15 to 25%, molasses contains a large carbohydrate fraction: its non-nitro-

* Corresponding Author Email: f.tendonkeng@univ-dschang.org.

genous extract represents about 65% of the fresh product consisting mainly of soluble carbohydrates (58.7%), essentially sucrose (34.6%), glucose (8.5%) and fructose (9.9%). On the other hand, it is deficient in nitrogen (4 to 6%) and has a high mineral content (12 to 16%). It is therefore an energetic food because of the importance of its carbohydrate constituents. It is rich in highly fermentable carbohydrate which stimulates appetite and improves the ingestion of roughage especially in ruminants. In cattle as well as in small ruminants, the synthesis of the results of numerous scientific studies shows that the inclusion of molasses (5 to 10%) in the diet, significantly improves the ingestion and digestibility of the main constituents of the fodder (dry matter, organic matter, proteins, NDF ...). Similarly, the introduction of molasses into the ration has repercussions on zootechnical performance and on the functioning of the gastrointestinal tract of monogastrics. This sub-product, in view of the growing interest in Cameroon's industrial landscape, is nowadays an important resource in terms of animal feed, soil fertilization in plantations and bioethanol production.

Keywords: Cameroon, molasses, chemical composition, animal feed

1. Introduction

Sugarcane "*Saccharum officinarum*" is a plant of the Poaceae family (grasses), grown mainly in the tropics and subtropics (Kapseu et al., 2014). There are several varieties with different characteristics that adapts to different agro-ecological zones and fluctuations in rainfall (Boli, 2007). According to Alfa (2005), these are characterized by their color, robustness, disease resistance, sugar content, etc. In Cameroon, sugar cane is produced by SOSUCAM (*Société Sucrière du Cameroun*). It was created in 1964 under the initiative of the Cameroonian government and since 2010, 72% of its capital is held by the French group SOMDIAA (*Société d'Organisation de Management et de Développement des Industries Alimentaires et Agricoles*). It covers an area of 23,000 ha of sugar cane plantations, and extends for almost 25 kilometers between M'Bandjock and N'Koteng (Boli, 2001).

With the exploitation of the varieties Co997 (56%) and B46364 (36%), the two sites produce annually 1.08 million tons (Zelakwa, 2013). Other varieties like B82333 and Fr81258 are in the experimental phase. With an extension of this plant's activities through a refinery in Mandjock, in 2012 a refined sugar production of 130,000 tons was observed. Local demand for sugar is divided between industrial consumers (brewing industries, fermentation companies) and households through wholesalers at around 100,000 tons. In addition, some of the production is destined for export to bordering countries like Chad. Cameroon occupies the leading position in sugar production in Central Africa. In addition, sugar byproducts (molasses) are processed to produce a variety of other products (food and pharmacy alcohol). "Sugar Valley" was created in Cameroon after the acquisition of CAMSUCO (Cameroon Sugar Company) by SOSUCAM. The sugar factory can cover the needs of a part of the population in energy by cogeneration. The by-products of sugar manufacturing such as molasses and bagasse are the raw materials of many other industries including alcohols, biofuels, yeast and animal feed.

Molasses is the principal by – product of the sugar industry. The term molasses specifically refers to the final effluent obtained in the separation of sucrose by repeated evaporation, crystallization and centrifugation of juice from sugar cane or sugar beets (Curtin, 1983). In general any liquid feed ingredient that contains sugar in excess of 43% is termed molasses.

Molasses is traditionally used in fermentation technologies to produce ethanol. Fermentation treatment of molasses to produce baker´s yeast or proteins is also tightly connected with ethanol production. Molasses produced in Cameroon is mainly used in the production of alcoholic drinks and also used as animal feeds. Molasses is applied in many food or non-food processes because of its high content in nitrogenous compounds, carbohydrates and its sweet taste. The use of molasses in road sector is as (i) dust palliative on the footpaths around sugar factories and (ii) to make molasses-based material for de-icing of roads.

2. Molasses Production in Cameroon

According to the report from a sugar corporation of Cameroon, the current aggregated national sugar production capacity from M'Bandjock and N'Koteng, sugar factories is estimated to be 130,000 tons of sugar per year. The amount of molasses produced from the extraction of sugar (produced at 3 to 4% of the tonnage of crushed canes) is estimated at 5,200 tons. This molasses is mainly used in the fertilization of plantations, in animal feed and in fermentation technologies to produce ethanol.

3. Use of Molasses in Animal Feeding

Molasses is suitable for inclusion in the diets of all ruminant livestock and can offer a very cost effective way to increase the palatability of feeds whilst contributing good levels of energy and protein. Initially the term molasses referred specifically to the final effluent obtained in the preparation of sucrose by repeated evaporation, crystallization and centrifugation of juice from sugarcane and from sugar beets. Today, several types of molasses are recognized and in general any liquid feed ingredient that contains in excess of 43% sugars is termed molasses. Molasses has been use in all livestock (cattle, swine, poultry…). The extent to which molasses has been used in animal feeds varies from small amount used, to eliminate dust and feed wastage, serving as the major source of dietary energy. It is a dark brown, viscous liquid produced as a co-product of the production of sugar. After dissolving sugar out at high temperature, the crystals of sugar settle out as the liquid cools leaving the molasses, much of which was traditionally mixed back with the pulped fibers to produce molasses sugar beet feed. Molasses, which is an excellent source of rapidly fermentescible energy, optimizes the use of ammonia produced by urea, and generates minerals (especially trace elements) (Chenost and Kayouli, 1997; Tendonkenget al., 2011) will be a good supplement.

3.1. Chemical Composition of Molasses and Nutritive Value

Molasses is a sticky dark by-product of processing sugar cane or sugar beets into sugar. Molasses can be a source of quick energy and an excellent source of minerals for farm animals (Tendonkeng et al., 2011; Tendonkenget al., 2014b; Matumuiniet al., 2014a; Tendonkenget al., 2018). It can also be common ingredient in animal feed (Tendonkeng et al., 2011; Fogang et al., 2012; Fogang et al., 2013). The calcium content of sugar cane molasses is high (up to one percent), whereas the phosphorus content is low. Cane molasses is also high in sodium, potassium, magnesium and sulphur. Beet molasses is higher in potassium and sodium but lower in calcium. Molasses also contains significant quantities of trace minerals such as copper, zinc, iron and manganese. Supplementing poor quality hay with molasses will increase feed intake and digestibility (Matumuini et al., 2013; Matumuini et al., 2014b; Lemoufouet et al., 2014a; Tendonkeng et al., 2018). Microbes in the rumen break down the sugars in molasses rapidly, which extensively causes a rapid release of energy that makes molasses very useful for balancing other feeds in the dairy diet all year round (Tendonkeng et al., 2015). Feeding molasses to farm animals will improve digestion of pastures/hay; increase milk production, help maintain body condition, appetite and result in less feed waste. Cane sugar, which has similar benefits to molasses, is an inexpensive alternative to use. The chemical composition of molasses is presented in Table 1.

Table 1. Chemical composition of molasses

Parameters	Normal molasses of sugar beet	Molasses of sugar cane
Dry matter (%)	73	73
Minerals matters (%DM)	13	14
Total Nitrogen Matters (%DM)	15	6
Totals sugars (%DM)	64	64
Calcium (g/Kg DM)	3.7	7.4
Phosphorous (g/Kg DM)	0.3	0.7
Potassium (g/Kg DM)	82	40

Source: Bernard et al. (1991).

The value of molasses as feed is based mostly on its sugar content (around 50 per cent). In comparison with carbohydrates in concentrated form, molasses contains small amount of protein, but it also provides a certain amount of non-protein, non-sugars which have some nutrient value, especially for ruminants. Table 2 present the nutritive value of molasses

Table 2. Nutritive value of molasses

Molasse of	UFL/ kg DM	UFV/ kg DM	PDIA g/kg DM	PDIN g/kg DM	PDIE g/kg DM	UFc/ kg DM	MADc g/kg DM
Sugar beet	1.03	1.04	0	84	71	1.06	83
Sugar cane	0.91	0.9	0	32	68	1.07	34

Source: Bernard et al. (1991).

Table 3. Nutrient composition of molasses (dry matter basis)

Nutrients	Content
Dry Matter	74%
Crude Protein	6.5%
Oil	Trace
NDF	Nil
Starch	Trace
Sugar	65%
ME (MJ/KG DM)	12.5

Sources: Senthilkumar et al. (2016).

In general, molasses should be added to feed when it is essential to compensate for an excess of protein according to his sugar content (Table 3).

Molasses has a high mineral content, but usually lacks adequate calcium and phosphorus. These must be taken into account when preparing mixed feeds and they should be supplied by suitable supplements (e.g., lime) or by a proper combination of feeding materials. Molasses is the uncrystallized sugar obtained after cooking cane juice during the manufacture of sugar in factories. Molasses contains about 25% water. It is a high-energy food containing uncrystallized sucrose (30% of the dry matter), reducing sugars (25% of the dry matter) and other carbohydrate substances. Molasses is particularly poor in nitrogen: 25 g of N per kilogram of raw molasses (INRA, 1988). As the fiber content of molasses is negligible, its consumption by ruminants must necessarily be accompanied by that of roughage. Many works has been conducted on the use of molasses by small ruminants in Cameroon (Boukila et al., 2013; Matumuini et al., 2014a, b; Lemoufouet et al., 2014a, b;Tendonkeng et al., 2014a, b, c; Tendonkeng et al., 2015; Tendonkeng et al., 2016; Tendonkeng et al., 2018). Rations based on molasses can reach growths of over 1000 g/day for cattle.

3.2. Benefits of the Use of Molasses

Molasses can reduce the dusty powdery nature of some finely ground feeds. In this role, it makes feed mixture more palatable and edible to livestock. Molasses can be added to replace missing sugar, trace minerals and help in fermentation in cases of low quality forage especially with low sugar levels.

3.2.1. Horses

Since horses are monogastrics herbivores, molasses can be used in their feed to improve both feed intake and digestibility. It can:

- Combines to reduce the dust in feed
- Increases palatability
- Reduces the ability of picky horses to sort through feed
- Prevents pregnancy toxemia

3.2.2. Feeding Regime for Horses

- 1-2 kg/450 kg of body weight two to three times a day

3.2.3. Cattle and Dairy

For this group of animals, molasses can:

- Provides sugar during early pasture growth
- Promotes animal health
- Increases milk solid production
- Increases diet density when intake is reduced before calving
- Improves milk let-down
- Improves digestion of fiber
- Helps reduce heat-related stress
- Helps growth and development of young stock
- Assists pregnancy rates (condenses calving patterns)

3.2.4. Feeding Regime for Cattle and Dairy

For a good efficiency of utilization of molasses in animal feed, it is necessary not to exceed some level of inclusion.

- Dry dairy cows: 500 g-1 kg/day
- Springing dairy cows: 700 g-2 kg/day
- Lactating dairy cows: 500 g-2 kg/day
- Adult beef cattle: 500 g/day
- Calves and heifers: 100-500 g/day

3.2.5. Feeding Regime for Sheep and Goats

- Lactating or in gestation: 100-200 g/day.

The high molasses feeds contain 30-40% molasses and the proportion by weight is limited to 2 or 3 parts of the permissible impregnated carriers. To a greater extent, molasses is used in various kinds of mixed feeds, which contain 5-15% and only occasionally more molasses. The standard table given in the German Feedstuff Law permits the following amounts of added molasses (Table 4).

Table 4. Amounts of added molasses according to the type of animal feeds

Animal feeds	Level of inclusion (%)
Calf feeds	5
Calf nutrition meals	5
Goat mixed foods	10
Dairy cattle feeds	15
Milk-producing feeds	15
Cattle fattening feeds	20
Horse mixed feeds	30

Sources: Adapted from Senthilkumar et al. (2016).

Molasses is suitable for inclusion in the diets of all ruminant livestock (Cattle, goat, sheep…) and even some herbivorous monogastrics (Horses, rabbit, Guinea pig…) and can offer a very cost effective way to increased palatability of feeds whilst contributing good levels of energy and protein. In dairy cows, ideal for complete diets added up to 3kg of molasses per head per day. Whereas, in beef cattle up to 10 per cent of molasses can be included in beef diets depending on the nature of other feeds, in the mixture and subsequent storage facilities for the finished ration. Similarly, up to 10 per cent of molasses can be safely included in young calve feed. But for herbivorous monogastrics it is necessary to reduce the quantity that should be added to the feed.

3.3. The Main Differences between Polygastric and Monogastric

Livestock are transformers of plant biomass. The digestive capacity of animals, the nature and quantity of end products are direct consequences of the anatomy of their digestive tract. The originality of digestion in polygastrics (including ruminants) is due to the activity of a multitude group of microorganisms, bacteria, protozoa and fungi, living in symbiosis with the animal. This microbial population lives in the reticulo-rumen (rumen) of ruminants, which are bio-fermenters producing proteins, volatile fatty acids. Thanks to these microbes, polygastric, unlike monogastric, can draw energy from fibers (celluloses and hemicelluloses). In addition, microbes allow ruminants to recover mineral nitrogen and synthesize amino acids as microbial proteins. The latter will then be digested in the gut of ruminants (Lemoufouet et al., 2014a, b, c; Tendonkeng et al., 2016; Tendonkeng et al., 2018).

Microbial proteins contribute between 40% and 60% of the animal's total needs according to the level of production. Ruminant benefits from the synthesis of vitamins B by microbes. Ruminants digests microbial proteins made up of practically all the amino acids essential to the body. Monogastrics are devoid of this microbial flora. Thus, at home, microbial digestion is often more limited and negligible. Monogastric therefore weakly digest celluloses and hemicelluloses. The main sources of energy are starch, sugars. Amino acids must necessarily be present in the diet in the form of proteins. The amino acid profile should be as close as possible to the animal's need.

Ruminants and pigs, therefore, do not have the same dietary constraints even though some feeds may be distributed to both "species." Ruminants can potentially combine a diet rich in fibrous products (grass, cane, bagasse), good performance and a state of well-being. Rations rich in low fiber products (industry concentrates, molasses, banana, potato, cassava, cereals) can cause digestive disorders and consequently penalizes performance. In contrast, rations rich in fibrous products do not allow pigs to perform well, in contrast to what is observed with low fibrous foods. Pig feeds must be high in starch and/or sugar and contain high quality protein. The cane is the main food (pork and ruminants) of the farm.

3.4. Molasses Supplementation for Monogastric Animal

Molasses and cane juice are the products most likely to be valued by pork. Cane juice can completely replace cereals in terms of energy input. For molasses the substitution is only partial. Unlike ruminants, pigs need good quality protein. Soybean meal is a reference food because it has an amino acid profile corresponding to the needs of pork. 200 to 400 g/day of soybean meal is required depending on the growth target. The meal may be replaced by another source of protein, but the candidates are relatively few in number. These are mostly agro-industrial products and some foliage (cassava, potato, erytrina, mulberry, madeira). It is necessary to favor the leaves richest in proteins and less lignified.

The common characteristic of molasses is that, it is a rich product in sugar and low in protein. In addition, the technological process of transformation increases the concentration of industrial impurities (non-digestible organic materials) and minerals. The raw energy content (GE) and digestibility of molasses energy are reduced compared to cereals. The relative weakness in GE is largely due to the importance of its mineral fraction which dilutes organic matter, and also because the raw energy content of sucrose, fructose and glucose is lower than that of starch. The metabolic use of nutrients derived from molasses are done with the same efficiency as for those derived from starch. Molasses would have a depressive effect on the use of proteins due to an increase in faecal nitrogen loss of endogenous origin. Molasses, however, remains a very available and inexpensive source of energy for cereals. However, the viscous consistency of molasses poses problems of texture and homogeneity of the complete ration. In addition there is a rate of incorporation not to be exceeded. High rates of molasses incorporation into diets result in digestive disorders leading to diarrhea. In pigs, this effect was attributed to the high content of molasses in minerals, particularly potassium, as well as the large unidentified organic fraction. Christon and Le Dividich (1978) set a 30% limit on the rate of incorporation of molasses into the ration of growing pork. Beyond this limit, the authors report a decline in growth performance. Molasses cannot represent more than 20% of the ration for piglet against 30% for the pregnant sow. The performance of the pigs will depend on the complementation used. It is also batter to know that Cane juice gives better performance than molasses. Work conducted in Cameroon and elsewhere on poultry feeding with molasses or cane juice to replace cereal starch gives different results depending on the animal species considered. Hens are bad users of liquid foods, unlike water fowl (ducks, geese). The productive life of hens is too short for them to adapt to a liquid diet. Similarly, their bill is not adapted to allow them to consume liquid foods. There are significant losses of liquid, splashes of sweet products on the plumage which are at the origin of problems of cannibalism. With hens, rations based on cane juice and molasses have always given much lower performance than the potential of animals. Ducks and geese are more suitable for consuming liquid feed. Performance equivalent to 80-90% of their genetic potential was recorded with cane juice (Senthilkumar et al., 2016). According to the same authors, the average growth of duck was 52 g/day for the rice ration, 46 g/day for the juice and 38 g/day for the molasses. The juice consumption averaged 436 g/day compared to 106 g/day for molasses.

3.5. Molasses Supplementation for Ruminants

In ruminants, the strategy will be to have the rumen microbes produce the maximum protein using non-protein nitrogen (urea, a fraction of the nitrogen in legume leaves). At the same time, it is necessary to provide good quality proteins, indigestible in the rumen, but digestible in the intestine. This is the case of soybean meal and nitrogen fraction of the leaves of plants rich in nitrogen (sweet potato, cassava, Madeira, legumes such as *Gliricidia* and *Leucaena*). Urea diluted in water can be mixed with bagasse, straw and mooring at the rate of 35 g of urea per kilogram of fresh bagasse. An energy supplement (molasses, banana, potato, rice flour, industrial concentrate ...) and a nitrogen supplement (soy or equivalent) are also essential. Urea may also be mixed with the whole crushed cane at the rate of 10 g of urea per kilogram of fresh material. Associated with urea according to the desired performances, it is necessary to bring 30 to 100 g of soybean meal per kilogram of fresh cane. Many sources of protein can be use in this case. Thus Matumuini et al. (2013); Matumuini et al. (2014a, b) have use with a good result *Tithonia diversifolia* as a nitrogen sources associated to molasses in small ruminants (goat and sheep) in Cameroon. Many others authors in Cameroon use nitrogen sources, urea or hens dropping in small ruminant diets (Boukila et al., 2013; Lemoufouet et al., 2014; Tendonkeng et al., 2014a, b, c; Lemoufouet et al., 2016; Tendonkeng et al., 2015; Tendonkeng et al., 2016; Tendonkeng et al., 2018). The addition of molasses in each case was used to improve the efficiency of Nitrogen utilization by animals.

The substitutes for soy are potentially numerous. These are all nitrogen-rich raw materials: leaves of Gliricidia, leucaena, erythrina, potato, sweet cassava or dried bitter cassava, peas, rice flour, other cakes, commercial foods. Legumes, some asteracea rich in crude protein, cassava leaves… can be also used in addition to molasses to improve the utilisation of very poor animals feeds like Straw, stoveer, hay… Molasses has also been included in multinutrient feed block for feeding small ruminant with good results by Tendonkeng et al. (2011), Fogang et al. (2012), Fogang et al. (2013) and Tendonkeng et al. (2014a).

On average, 300g of commercial feed containing 15% MAT is needed to replace 100g of soybean meal. It takes 1 to 2 kg of fresh foliage rich in nitrogen to replace 100 g of soybean meal.

3.5.1. Molasses in Small Ruminant Feeding

Molasses can be used, associated to different protein sources to improve feed intake and digestibility. Thus many studies have been conducted in Cameroon using some legumes forages, hens dropping or urea as nitrogen sources associated to different levels of molasses to improve feed intake and digestibility in small ruminants (Lemoufouet et al., 2014a, b; Tendonkeng et al., 2014a, b, c; Lemoufouet et al., 2016; Tendonkeng et al., 2015; Tendonkeng et al., 2016; Tendonkeng et al., 2018). The study of the effect of the level of incorporation of molasses (0, 5 and 10%) and the source of the ruminal fluid (Djalonké lamp and the African Dwarf goat) on *in vitro* digestibility of rice straw treated with 5% urea was then conducted in Cameroon by Tendonkeng et al. (2016). This study showed that, gas production was not affected by the addition of molasses to rice straw treated with 5% urea. On the other hand, dry matter mineralized increased significantly with the level of incorporation of molasses regardless of the source of rumen fluid. With the sheep rumen fluid, 5% molasses ration helped to get the highest gas production (31.3 ml/500mg). Generally goat rumen fluid gave the best results whatever the ration considered. When maize stover were associated with *Tithonia diversifolia* leaves treated with

5 or 10% molasses, feed intake and digestibility of maize stover was improved in goat (Matumuini et al., 2014b).

The evaluation of the effect of molasses level on the ingestion and *in vivo* digestibility of rice straw treated with 5% urea was also carried out on small ruminant (Tendonkeng et al., 2018). The results of this study show that, the addition of molasses enhances the digestibility of organic matter and digestible crude protein of rice straw treated with 5% urea. Dry matter, organic matter and NDF intake of the ration without molasses was higher than that of others rations. The highest apparent dry matter (DM), organic matter (OM) and NDF digestibility were obtained with the ration containing 5% molasses. The apparent nitrogen digestibility of this ration was significantly higher than that of ration without molasses and comparable to that of the 10% molasses ration. These results shows that the inclusion of molasses in rice straw treated with 5% urea does not significantly affect the ingestion and digestibility of DM, OM and NDF, but significantly improves digestibility of nitrogen.

Simillar study on intake and *in vivo* digestibility of maize stover treated at 5% urea associated with 0, 5 and 10% molasses in West African dwarf goats was conducted (Tendonkeng et al., 2014b, c). Results of this study shows that the addition of molasses increased the level of organic matter digestibility of maize stover treated at 5% urea. The dry matter and organic matter intake at 0 and 5% molasses diet were comparable and significantly higher than that of 10% molasses diet. The apparent digestibilities of dry matter and organic matter of 0 and 10% molasses diets were comparable, and significantly lower than that of 5% molasses diet. The digestibility of the cells wall (NDF) of 5% diet were significantly higher than that of 10% molasses diet. The addition of molasses significantly increased digestibility of nitrogen of 5% molasses diet. These results shows that, the addition of 5% molasses in maize stover treated at 5% urea increases intake and digestibility.

The study of the effect of the level of incorporation of molasses and the source of the ruminal fluid on *in vitro* digestibility of maize stover treated with 5% urea or with 28% of poultry dropping was conducted by Lemoufouet et al. (2016) showed that, the addition of molasses in feed significantly improved the level of gas production after 24h of maize stover treated with 5% urea regardless the ruminal fluid (from goat or sheep). The same tendency was observed with this parameter when maize stovers were treated with 28% of poultry dropping associated at various levels of molasses addition. The dry matter mineralized were significantly influenced by the addition of molasses to maize stover treated with 5% urea, as well with the sheep and goat rumen fluid. Generally, goat rumen fluid gave the best results. As previous research, it appears that 5% rate of molasses seem to be a good level of inclusion in small ruminant feeds. This is confirmed with the trial conducted by Matumuini et al. (2014a) and Tendonkeng et al. (2018).

The study of the effect of the level of incorporation of molasses and the source of the ruminal fluid on *in vitro* digestibility of rice straw treated with 5% urea was conducted in Cameroun by Tendonkeng et al. (2018) using two sources of ruminal liquid (from sheep and goat) and three levels of molasses (0, 5 and 10%). This study showed that gas production was not affected by the addition of molasses to rice straw treated with 5% urea. On the other hand, the dry matter mineralized increased significantly with the level of incorporation of molasses regardless of the source of rumen fluid. With the sheep rumen fluid, 5% ration induced the highest gas production (31.3 ml/500mg).

At the end of the study, which examined the effect of molasses level on intake and *in vivo* digestibility of maize stover treated with 28% hens dropping in small ruminants, it was found that the inclusion of molasses on maize stover treated with 28% hens droppings increased nutrient content (dry matter organic matter, total sugar). The addition of 5% molasses to 28% chicken droppings resulted in a significantly higher intake of dry matter (DM), organic matter (OM), and nitrogen in the Djallonke sheep, while goat, ingestion of these nutrients was higher with 10% diet without any significant difference being observed. The inclusion of molasses in 28% of chicken manure treated with treated stubble significantly improved the digestibility of dry matter and organic matter in sheep, while in goats, digestibility of these nutrients were comparable. The addition of 5% molasses allowed better digestion of DM and nitrogen of maize stover by goats. Although the results of this study are satisfactory with 5% molasses, it would be desirable to further this work by evaluating the effect of 5% molasses on growth performance in small ruminants. Also, the study of the quality of the products (milk and meat) of these animals nourished with this ration would be advised to check the quality of the products intended for human consumption (Lemoufouet et al., 2014a, b). All these works shows that, molasses is a very good energy source, suitable for the transformation of very poor farm residues.

3.5.2. Performance of Beef Cattle Fed Molasses

3.5.2.1. Diet Intake

One of the most important characteristics of feedstuff is its influence on diet intake, because of the close positive relationship on intake to animal performance and production efficiency. The first and most recognized benefit of feeding molasses to cattle has been its ability to improve diet palatability. Datas shows that, in most cases, the addition of up to 10% molasses to both roughage and concentrate diets improved daily dry matter intake. Although this response has usually been attributed to improved taste or reduced diet dustiness, the low levels of molasses to increase fiber digestibility and microbial activity may be responsible (Lemoufouet et al., 2015; Tendonkeng et al., 2016; Tendonkeng et al., 2018).

Less known is the relationship between diets containing moderate to high levels of molasses and feed intake, and how this relationship relates to the established mechanisms known to control diet intake (Balch and Campling, 1962; Conrad et al., 1964; Conrad, 1966). Silvestre et al. (1978) conducted a study in which growing bulls were fed sugar cane based diets containing up to 41% cane molasses and found that dry matter intake increased linearly with increasing levels of molasses. A similar response was reported by James (1973) and Toranzos et al. (1975) when 43% or 30% cane molasses was added to chopped sugar cane or sorghum silage based diets, respectively. Bond and Rumsey (1973) and Delgado et al. (1978) also reported that, the *ad libitum* supplementation of hay or fresh forage based diets with cane molasses (39 and 23% of diet dry matter, respectively) substantially increased in daily dry matter intake. Brannon et al. (1954) reported that the daily dry matter intake by grazing steers was increased from 6.0 kg to 6.7 kg with the *ad libitum* supplementation of 1.0 to 1.5 kg per day of cane molasses. Comparing different supplements, Bohman et al. (1954) and Merrill et al. (1959) observed that heifers supplemented with molasses consumed more forage dry matter and total dry matter than heifers supplemented with a similar quantity of corn. In contrast, King et al. (1960) noted no difference in the intake of oat hay by heifers fed either molasses or corn supplements. The above data on molasses feeding support the general concept that dry matter intake by cattle increases with increasing concentrations of digestible nutrients in a forage based diet.

At the opposite extreme, that is, diets with high levels of molasses and low levels of forage, the data available are very limited. Elias et al. (1969) fed growing cattle with restricted quantities of forage and cane molasses *ad libitum*, such that, the diets contain 75 to 90% molasses-protein supplement. Data showed that, increasing levels of molasses, or decreasing levels of forage, linearly decreased daily dry matter intake. It is also interesting that daily weight gain of cattle fed diets containing different levels of molasses were similar, suggesting that available energy intake was similar in steers fed with each diet. In a study with *ad libitum* molasses feeding, Martin et al. (1968) found daily dry matter intake of growing bulls fed on restricted forage diet which contained 85% molasses was substantially less than the intake by bulls fed *ad libitum* with forage diet which contained 29% molasses. Again, these intake data of diets containing high levels of molasses support the general concepts relating to the control of diet intake that is, physiological factors limit intake of highly digestible diet even when molasses is the concentrate energy source. Studies in which molasses is added to or substituted for concentrate ingredients in high energy fattening diets also give evidence as to the effect of molasses on feed intake. Lofgreen and Otagaki (1960a) reported that the addition of 10% cane molasses to a relatively fibrous fattening diet fed to steers increased dry matter intake, but further additions of 25 or 40% molasses drastically reduced intake. This curvilinear relationship between the level of molasses in the diet and dry matter intake tended to be confirmed by Heinemann and Hanks (1977) when 0, 10 and 20% molasses was fed *ad libitum* with a barley based fattening diet. O'Mary et al. (1959) also reported a much lower intake of dry matter by steers fed diet containing 47% cane molasses than that of steers fed on diet containing 54% corn, but this result may have been influenced by the use of different roughage ingredients, cottonseed hulls (CSH) and alfalfa hay respectively, in the two dietary treatments.

A number of studies (Bray et al., 1945; Riggs and Blankenship, 1955; Brown, 1962; 1967; Campbell et al., 1970) have shown that increasing levels of molasses of up to 48% of feedlot diets has little effect on dry matter intake by growing or fattening cattle. Bray et al. (1945) did observe that when cane molasses was substituted for 10 to 15% of corn grain or other dry ingredients in the concentrate ration, steers consumed more hay or silage which were offered *ad libitum* as roughage component, but higher substitutions of molasses did not appear to encourage a further increase in the intake of roughage.

In a second study, Lofgreen (1965) found that the substitution of cane molasses at 5, 10, 15 and 20% of barley in a fattening diet resulted in a slight linear increase in dry matter intake by heifers, but intake of all diets containing molasses was below that of heifers fed on control diet. Feeding studies by Lishman (1967), Van Niekerk and Voges (1976) and Kargaard and Van Niekerk (1977) showed that the substitution of cane molasses for up to 22 to 30% of corn meal in steer finishing diets resulted in increases in dry matter intake. In a feeding study involving 15 separate feedlot trials, Baker (1954) found that the substitution of citrus molasses for up to 50% of ground ear corn in finishing diets resulted in an increase in dry matter intake over the controls, but in a subsequent study (Baker, 1955a) there appeared to be a negative relationship between the level of citrus molasses in the diet and dry matter intake by fattening steers. Gaili and Ahmed (1980) reported a much higher intake of dry matter by growing cattle fed diets containing 25 and 50% cane molasses than that of cattle fed a 45% sorghum grain diet.

In two studies involving the substitution of raw sugar for up to 40 and 48% of corn meal in steer fattening diets, Beardsley et al. (1971) and Olbrich and Wayman (1972) noted little effect of dietary treatment on dry matter intake indicating that the effects of sucrose and starch were similar. However, when diets contained different combinations of cane molasses and raw sugar, increasing levels of molasses consistently increased dry matter intake by fattening steers (Olbrich and Wayman, 1972).

Although the types of molasses are somewhat different in composition, studies (Baker, 1955a, b; Riggs and Blankenship, 1955; Kirk et al., 1966; Crawford et al., 1978; Cooper et al., 1978) that have made direct comparisons between diets containing either cane, corn, citrus or wood molasses have not demonstrated consistent differences in terms of dry matter intake. However, the type of feed ingredients with which molasses is combined does influence intake. Both Brown (1962; 1967) and Salais et al. (1977) found that sugar cane roughage feeds fed in combination with cane molasses could be quite detrimental to dry matter intake relative to other roughage sources. Baker (1966) reported that the addition of 15% molasses to a ground ear corn diet increased dry matter intake, but a similar quantity of molasses added to a shelled corn diet did not affect intake.

3.5.2.2. Molasses in Fattening Diets

A number of studies (Bray et al., 1945; Lofgreen and Otagakit, 1960a; Bradley et al., 1966; Lishman, 1967; Brown et al., 1967; Copper et al., 1978) have shown a very obvious advantage in rate of gain and/or dry matter utilization from the addition of 2 to 10% cane molasses to concentrate diets fed to finishing cattle. Only in three studies reviewed did the feeding of up to 10% cane molasses did not result in an improvement in animal performance in comparison to the control treatment (Van Niekerk and Voges, 1976; Kargaard and Van Niekerk, 1977; Hinemann and Hanks, 1977). In either of the above studies carcass quality was not measurably affected by the feeding of low levels of molasses. The above data strongly support the conclusions of the basic studies that, the addition of up to 10% molasses to finishing diets stimulate microbial activity, the digestibility of energy and fiber, and nitrogen utilization (Potter et al., 1971; Hatch and Beeson, 1972; Crawford et al., 1978).

Basic studies have suggested that the addition of low levels of wood molasses might protect dietary protein from bacterial attack thereby increasing by-pass protein and dietary protein utilization. However, practical feedlot studies (Riggs and Blankenship, 1955; Cooper et al., 1978; Crawford et al., 1978) have not shown a consistent difference in the performance of fattening cattle fed low levels of either wood or cane molasses.

The response of fattening steers to the feeding of molasses may be related to the type of diet in which it is substituted. Baker (1966) found that the addition of 15% citrus molasses to ground shell corn diet reduced, rate of gain and dry matter utilization by fattening steers by 13 and 10%, respectively, but a similar quantity of citrus molasses added to ground ear corn diet increased rate of gain to a level which was equal to that obtained with the ground shelled corn diet, but did not influence dry matter utilization which was 15% lower than that obtained with the ground shelled corn diet. Lishman (1967) reported that the substitution of cane molasses with 20 and 30% of corn meal in a corn silage diet increased rate of gain by fattening steers by 15%, but it did not influence the efficiency of dry matter utilization. In a series of 15 feeding trials, Baker (1954) found that the substitution of citrus molasses with 22 and 37% of ground ear corn in steer finishing diets improved rate of gain by 28% and the efficiency of dry matter utilization by 11 to 15%. The performance of steers fed diets containing 50% citrus molasses was similar to that of steers fed the control diet, but in a subsequent study (Baker, 1955a), steers fed diet containing 51% molasses as a substitute for ground ear corn had a similar gain and were 6% more efficient than steers fed the control diet.

Several studies have compared the feeding value of different types of molasses in steer finishing diets. Baker (1955a) reported that citrus molasses had a higher feeding value than blackstrap molasses when both were fed with 40% ground ear corn based diet. Steers fed on ground ear corn diet containing 20% blackstrap molasses performed better than steers fed diet containing 20% standard cane molasses (Baker, 1955b). Kirk et al. (1966) observed little difference in the performance of fattening cattle fed diets containing 29% of either blackstrap or citrus molasses. Riggs and Blankenship (1955) fed diets containing 13 and 26% molasses of four different types and reported that fattening cattle performed best on diets containing blackstrap molasses, followed by corn, citrus and wood molasses respectively.

Two studies have shown that, raw sugar was superior to corn meal in energy ingredient in steer finishing diets (Beardsley et al., 1971; Olbrich and Wayman, 1972). The substitution of raw sugar for up to 48% of corn meal did not affect the rate of gain by fattening steers, but increasing levels of raw sugar tended to improve the efficiency of dry matter utilization by about 10%. These data indicate that, sugars, the principle component of cane molasses, did not adversely affect the performance of finishing cattle when included into their diets, at moderate to high levels.

In the early 1950's Wayman and co-workers, in Hawaii, initiated a series of studies to develop steer fattening diets using cane molasses as the major energy source. In the initial studies (Wayman et al., 1952; 1953; 1954), three problems were associated with the feeding of high-molasses diets:

1) the adaptation of cattle to utilize molasses based diets was very critical and should be done over a period of several weeks,
2) the feeding of fresh-chopped forage was essential, especially during the adaptation period, and
3) the level of sugarcane bagasse, the dry roughage source, should be limited to less than 10% of the diet and the fresh-chopped forage should be limited by restricted feeding to about 5 to 7% of diet (dry matter basis) after the cattle have adapted. Further investigations found that dehydrated legume forage could be completely substituted for fresh forage after an initial two week period during which fresh forage was fed.

In one study (Olbrich and Wayman, 1972) it was shown that the rate of gain and efficiency of dry matter utilization of steers fed a 55% cane molasses diet was 83 and 72%, respectively, of that by steers fed a 60% corn meal diet. The utilization of total digestible nutrients in the molasses diet was 78% of that in the corn meal diet.

More investigations to develop high-cane molasses diets for the commercial fattening of beef cattle were conducted by Preston et al. (1967a;

1969; 1970a) in Cuba. They confirmed, the conclusions reached by Hawaiian workers, that cattle had to be slowly adapted to molasses based diets through the initial feeding of fresh-chopped forage, after which fresh forage should be limited to 1.5 kg per 100 kg of body weight or approximately 10 to 15% of the diet dry matter. It was also found that high intake of molasses and relatively good animal performance could also be obtained by restricted grazing (Morciego et al., 1970). The importance of restricted forage feeding on the performance of growing bulls was demonstrated in commercial feeding operations (Munoz et al., 1970).

An additional factor introduced into this high-molasses feeding system was the utilization of large quantities of non-protein nitrogen avoiding the use of natural protein concentrates if possible. However, in comparison to sorghum grain diet, bulls fed with molasses-urea diet performed poorly, particularly in terms of dry matter utilization (Preston et al., 1967b). Further investigation showed that some natural protein in the molasses-urea diet was essential and it was recommended that fattening bulls receive 140 gm of fish meal daily per 100 kg of liveweight because of its ability to provide by-pass protein (Preston, 1969; 1972). The performance of bulls fed fresh forage meal diet on an experimental (Preston et al., 1970b) and commercial feedlot basis (Munoz et al., 1970). The only abnormal problem encountered with this feeding system was a high incidence of molasses toxicity in the feedlot program. The incidence of this problem was much lower in the restricted grazing program (Morciego et al., 1970).

Other investigators have also tested high-molasses feeding system and made comparisons to the performance of cattle fed more conventional diets. In Kenya, Creek et al. (1974) reported that, gain and dry matter conversions by steers fed diet containing 53% cane molasses were 25 to 30% lower than that of steers fed 58% corn silage diet. They, too, encountered a high incidence of molasses toxicity (18%) which was completely corrected by replacing corn silage with straw, feeding some cereal grain, using a moderate level of urea, and using molasses minimally diluted with water. Molina (1977) fed growing calves to a constant weight and found that those fed cane molasses-urea-fish meal and restricted forage gained 30% slower than calves fed sorghum grain, but 36% faster than calves fed fresh napier grass forage (feed intake data were not presented). In Sudan, Gaili and Ahmed (1980) found that crossbred bulls fed diets containing 25 and 50% cane molasses gained similarly to bulls fed 45% sorghum grain, 33% wheat bran diet, but were respectively 12 and 32% less efficient in converting dry matter to gain. Again, a high incidence of molasses toxicity (23%) was encountered with cattle fed diets containing molasses.

As a summary, the following conclusions were drawn from the literature studied on the feeding of molasses in beef cattle fattening diets:

1) The addition of less than 10% molasses to concentrate fattening diets has a stimulating effect on animal performance, improving feed intake, rate of gain and/or feed utilization.
2) The feeding of fattening diets containing 20 to 40% molasses reduces the rate of gain and/or feed efficiency but to a degree that is explained by the energy content of molasses relative to the energy content of ingredients for which it is substituted. The majority of the feeding data do not suggest, that the energy efficiency of molasses declines, when its level in the diet exceeds 10 to 20% level.
3) At moderate levels, molasses appears to be better utilized when it is fed with certain concentrate feeds such as ground ear corn. This, and other information, suggest that molasses combines best with certain levels and kinds of fiber in a complete diet.
4) Several studies have demonstrated that high levels of molasses can be formulated into diets for fattening cattle. The success of this feeding system is very sensitive to feed management practices, particularly during the animal adaptation period, and diet composition in general, the production data suggest that the metabolizable nutrients of diets containing high levels of molasses are utilized less efficiently than those of diets formulated from more conventional concentrates. The economics of production is the most important factor and may indicate that feeding high-molasses diets is justified in many parts of the world.

3.5.2.3. Molasses in Forage Diets for Growing Cattle

An often stated beneficial use of molasses is its addition to diets based on low quality forages or roughages to improve palatability and provide a readily available source of energy. But, it must be remembered that molasses contains little crude protein and for it, or the diet to which molasses is added, to be efficiently utilized a source of supplemental crude protein is theoretically required. This concept was clearly demonstrated by Delgado et al. (1978) with yearling bulls fed fresh pangola grass which contained 4.7% crude protein. Supplementation with 1.4 kg of cane molasses actually reduced rate of gain, although total dry matter intake was substantially increased. Adding urea to molasses markedly increased rate of gain and the addition of sunflower seed meal resulted in an even better rate of gain. Toranzos et al. (1975) also reported a very positive response in rate of gain by steers fed sorghum silage diet when supplemented with 3 kg of molasses-urea mixture. Bond and Rumsey (1973) found that weaned calves or yearling steers fed hay diets which contained 9.4 and 4.3% crude protein, respectively, did not respond to molasses supplementation, but neither was there a response to a molasses-urea supplement.

James (1973) conducted a study with derinded sugar cane, cane tops and pangolagrass diets in which a urea containing (60% of the N) protein-mineral-vitamin supplement was fed in all experimental treatments. Supplementation with 3.8 kg of molasses dry matter (33% of diet) increased rate of gain by 10 to 15%. However, supplementation with 3.3 kg of corn grain increased rate of gain by 30%. Several review studies, did not use a negative control (forage alone) or the molasses supplement was fortified with urea or natural protein. These studies demonstrated that molasses was equal to corn grain as an energy supplement in forage based diets fed to growing heifers, if plant protein provided the supplemental nitrogen (Bohman et al., 1954; Davis et al., 1955; Merrill et al., 1959; King et al., 1960). In studies where urea provided the nitrogen source in molasses supplement a lower rate of gain was obtained.

Silvestre et al. (1978) fed growing bulls sugar cane based diets containing 0, 19, 32 and 41% cane molasses and found that the addition of 19% molasses improved rate of gain, but the animals did not significantly respond to further increases in the percentage of molasses in the diet even though there was a linear increase in dry matter intake. Brown (1962; 1967) also noted little difference in rate of gain or dry matter intake by steers fed bagasse diets containing from 25 to 40% cane molasses. But, in a second study there was an increase in both dry matter intake and rate of gain by steers fed bagasse diets as the level of molasses increased from 35 to 50%. At the opposite extreme, Elias et al. (1969) reported that increasing levels of forage in diets containing 61 to 70% molasses increased total dry matter intake by fattening bulls, and slightly increased rate of gain.

The response to molasses supplementation appears to be related to the roughage with which it is fed. Brown (1962; 1967) reported a much lower rate of gain by steers fed bagasse diet containing 20% cane molasses than that of steers fed grass hay, rice straw or cottonseed hull diets containing 20% molasses. The best rate of gain was obtained with rice straw diet. However, when steers were fed these same roughages with 40% molasses there was little difference in the performance of animals fed with different roughage sources. With growing bulls fed diets containing 80% molasses, Salais et al. (1977) noted a much lower rate of gain when either sugar cane or cane tops were used as a roughage source than when either Bermuda grass or a mixture of Bermuda grass and a legume forage was provided. Forages were fed as fresh chopped material.

From the above data the following conclusions can be drawn relative to the addition of molasses to forage or roughage diets:

1) A nitrogen supplement should be provided with molasses when added to low quality forage diets, and natural protein is superior to non-protein nitrogen sources. This is also applicable when feeding small ruminant with molasses as supplement (Tendonkeng et al., 2015; Tendonkeng et al., 2016; Tendonkeng et al., 2018).
2) Molasses supplementation will usually result in a lower intake of forage dry matter but an increased intake of total dry matter.
3) The benefits in rate of gain obtained with molasses supplementation are disproportionately lowers relative to increased obtained in total

dry matter intake. This may result from poorer utilization of forage dry matter.

4) The response to molasses feeding appears to be related to the forage or roughage with which it is fed.

3.5.2.4. Molasses Supplementation of Growing Cattle on Pasture

Molasses is most often fed to growing cattle as a supplement to pasture. Many studies indicated that the response to molasses supplementation was low whether fed during the dry season when the quantity and quality were limited or during the rainy season when forage was abundant. Even in studies conducted on mineral soil where forage availability or quality were not extremely limited, growing cattle have shown an inconsistent response to molasses supplementation. Delgado et al. (1975) and Vilela et al. (1976) presented evidence that grazing cattle might respond best to molasses-urea feeding when forage availability was extremely low, as would occur with dry season pastures. This response was not confirmed by Copeman et al. (1977) with steers grazed under similar conditions and fed a molasses-natural protein-urea supplement in northern Australia. Porres (1971) and Martin and Alfonso (1978) observed a poorer response by grazing cattle to molasses or molasses-urea supplementation during the dry season than during the wet season.

The response of growing cattle to molasses supplementation and its relationship to forage availability has been best demonstrated in studies involving varying stocking rates. Chapman (1965) showed that the response of grazing cattle to molasses feeding was much higher (.16 vs .09 kg per day) on heavily stocked pastures than when the molasses fed group was compared to an unsupplemented group of steers grazed at a lower stocking rate. Hart et al. (1971) graphically demonstrated this relationship between stocking rate and response to molasses supplementation. By progressively increasing the stocking rate of steers grazing orchard grass pasture, the response to the feeding of 4 kg per steer per day of cane molasses was increased to approximately 0.2 kg daily of additional gain, which appeared to be the maximum response obtainable. This maximum response by grazing steers to molasses feeding would be supported by literature in general.

The above data bring out two points relative to supplementing grazing steers with molasses. Firstly, there is a slight additive response in rate of gain obtained with molasses feeding, probably through an increase in total daily dry matter intake as was demonstrated with grazing steers by Brannon et al. (1954). Secondly, molasses feeding to grazing cattle substitutes in part for intake of forage as was shown in the previous section which discussed the feeding of molasses in forage or roughage diets. Thus, the feeding of molasses to grazing cattle also extends the availability of pasture forage or increases stocking rate. In fact, this latter point was accurately demonstrated by Mott et al. (1967). Their data showed that, the feeding of 2 kg per day of cane molasses to steers grazing guinea grass pasture increased daily gain up to .07 kg per day (10 to 15%) and increased stocking rate up to 0.5 animals per hectare (15 to 20%). If molasses is fed to growing cattle grazing pasture only during the dry or winter seasons perhaps, its benefits should be viewed solely from the standpoint of an increased stocking rate. In a study similar to that conducted by Mott et al. (1967), involving dry concentrate supplement fed only during the dry season, Bisschoff et al. (1967) found that the increased gain of growing steers obtained from supplementation during the dry season were lost during the subsequent wet season when forage was abundantly available.

Several studies have made direct comparisons between molasses and other energy feeds as supplements to growing cattle on pasture (Carlo et al., 1972; Holder, 1972).In general, the results show that corn and other concentrate feeds were superior to molasses in terms of increasing rate of gain.

Molasses is often mixed with additives for use as a supplement to cattle grazing pasture. The most important of these is non-protein nitrogen compound, usually urea. Several studies have shown an advantage of adding urea to molasses (Lemoufouet et al., 2014a, b; Tendonkeng et al., 2014b, c; Tendonkeng et al., 2018), but others did not indicate an advantage in rate of gain with the addition of urea to a molasses supplement (Mott et al., 1967) or showed that cattle were less responsive to molasses-urea than to other energy-protein supplements (Holder, 1972).

In addition to being mixed with non-protein nitrogen compounds, molasses has been used as a carrier of many minerals, vitamins, growth stimulants and medicinals (Chapman and Pace, 1974). In some cases, the feeding of these additives with molasses appeared to be superior to other methods of administration. For example, Copeman et al. (1977) reported a response by growing cattle to copper and cobalt supplementation when added to molasses as compared to when these elements were administered through injections to the animal. These data also suggest that certain additives may be beneficial to the utilization of molasses.

To summarize the above discussion the following conclusions are offered:

1) The intake of molasses by growing cattle on pasture is relatively low (2 to 4 kg per day), thus molasses itself is not a very palatable feed.
2) Growing cattle fed molasses supplement will gain an additional 0.1 to 0.2 kg per day with a relatively poor efficiency in terms of additional gain per unit of molasses consumed.
3) Grazing cattle fed molasses tend to eat less forage, thus it should be fed only in periods when pasture forage is limited. This would maintain higher stocking rates during these periods, which would allow more efficient utilization of forage available during the summer or wet season growth periods. Any benefits obtained in increased gains from molasses supplementation during the dry or winter season may be lost during subsequent periods when forage is more available.
4) Molasses supplementation to grazing animals is the area of production that would benefit from additional research. Many questions remain unanswered relative to the use of non- protein nitrogen and other additives that could improve the utilization of molasses and the total supplemented diet.

3.5.2.5. Molasses for Brood Cows

In recent years, molasses has been increasingly used as a supplement for brood cows. Crude protein requirements are more than adequate under the conditions of this study. Treatments included an unsupplemented control, the feeding of 2.3 kg daily of molasses during a 130-day winter period which included the breeding season, and the feeding of 2.3 kg daily of molasses year-round. Six breed groups were used, which included straightbred Angus, Brahman and Hereford cattle and the three possible two-way cross-breeds (cows and calves) of these three breeds. There was a definite breed difference with respect to straight-bred vs. cross-bred cattle. The winter feeding of about 300 kg of molasses to straight-bred cows increased both cow reproduction and calf survival and weaning rate, which resulted in a 26 kg increase in annual calf production per cow. Winter/dry season supplementation of cross-bred cows primarily increased calf weaning weight, and annual calf. Seasonal and year-round molasses feeding also increased cow weights by 15 and 20 kg, respectively, which would have economic implications in terms of cull cow sales.

Many results suggest that range cows respond best to year-round feeding of a molasses-urea mixture which includes, the spring breeding season. It was interesting that with year-round feeding cows consumed about 2 kg daily of liquid supplement during the summer as compared to only 0.9 kg during the winter months. The question presented is to know what extent the benefits derived from year-round supplementation were due to the additional energy provided by molasses in comparison to the additional nitrogen.

In a series of one-year feeding experiments, Rush and Totusek (1976) found that grazing brood cows fed dry supplements containing natural protein or urea during a 140-day winter/dry season period tended to lose less weight than cows fed on liquid supplement containing urea. Also, cows fed 1.6 kg daily on liquid supplement containing urea lost less weight than those fed 3.1 kg of cane molasses. However, most of the cows that lost weight during the winter gained more weight during the subsequent summer period, and the birth and weaning weights of calves were similar regardless of the winter supplementation regime. In a 118-day wintering trial Bond and Rumsey (1973) found that non-lactating beef cows fed timothy hay containing 4.3% crude protein lost less weight than cows fed hay and 2.1 kg of cane molasses or cane molasses-urea daily.

Brown (1962; 1967) evaluated different roughage sources in 40%, cane molasses supplements fed to brood cows on open range. During a 41-day winter feeding period, cows fed supplements containing either bagasse or cottonseed hulls lost considerably more weight than cows fed supplements containing grass hay or rice straw (roughage was 27.5% of supplement). These data support those results previously discussed relative to the influence of roughage source on the response of growing cattle to diets containing molasses.

Molasses were used, in Cameroon, to fatten cattle in an experiment on three types of animals - zebu steers, old cows and yearling bulls. In the three cases, the rations consisted of limited quantities of hay, a protein supplement of cotton seed cake and molasses *ad libitum* (Lhoste, 1980). The result are summarized in Table 5.

Table 5. Mean weight gains and energy conversion coefficients appear as follows

Animals	Average daily gain	Energy conversion coefficient (UF/kg)
Steers	915 g/day	7.9
Yearling bulls	794 g/day	7.9
Old cows	1480 g/day	10.6

High average daily weight gain in recent trials seems to favour short factening periods (about 2 months). The Adamawa zebu appears to be a good converter of this by-product which abounds in Cameroon. These trials have established the technical bases molasses use in feed lots but additional studies need to be done in the sugar producing area. The amount of nitrogen retained increased with the increasing level of molasses in the diet, with no significant difference observed. These results confirm that, the addition of molasses to rations based on roughages improves the digestibility of nitrogen (Swanson et al., 2004). Indeed, microorganisms would use the energy provided by molasses and ammonia produced by the fermentation of roughages to synthesize microbial proteins. In a similar study on small ruminant, Tendonkeng et al. (2018) concluded that, the addition of 5% molasses significantly increased apparent digestibility of nitrogen in very poorly treated roughages with 5% urea. This is explained by the fact that molasses provides the energy and urea carbon skeleton necessary for the microorganisms of the rumen to improve the degradation of the ration and the passage of digesta from the rumen to the abomasum, resulting in improved digestibility of nutrients (Chenost and Kayouli, 1997; Matumuini et al., 2014a, b). At the end of his study, he observed that cell wall (NDF) and total carbohydrate contents decreased with the increasing level of molasses in the rations. On the other hand, the contents of total crude protein, digestible crude protein and organic matter digestibility (OMD) increased with the level of addition of molasses in the rations. The addition of 5% molasses to rice straw treated with 5% urea helped to obtain the best DM, OM and NDF digestibility. However, the addition of different levels of molasses to rice straw treated with 5% urea did not significantly influence the intake and digestibility of DM, OM and Cell walls (NDF). The addition of molasses to rice straw treated with 5% urea significantly improved the apparent digestibility of nitrogen in West African dwarf goats. Although he believes that the results of this study are satisfactory with 5% molasses, but it would be desirable to evaluate the effect of adding 5% to urea-treated rice straw on the production performance of the West African dwarf goat. A similar study could be considered in Djallonké sheep to assess its suitability for using this crop residue. Based on this suggestion and many others aspect lacking in this work, it appears that many aspects of the topic remains virgin and may be open to many others studies.

4. Other Uses of Molasses

4.1. Soil Improvement and Plant Fertilisation

Many studies have focused on the use of molasses in improving the physico-chemical properties of the soil over the world. The study of Ravi et al. (2015) with the aims to find the effectiveness of uses of unconventional liquid soil stabilizer i.e., molasses for improving the shear strength and CBR value of two types of fine grained soils showed that, with the use of liquid stabilizer, there was appreciable increment in unconfined compressive strength and CBR value of both soils. The unconfined compressive strength of soil increased with increment ratio in range of 1.57-2.01 for both types of soils. CBR value of soils had increment ratio of 2 – 3.5 in both types of soils. The values increased with increased in treatment duration. Based on this study, optimum amount of liquid additive to be added to respective soil types for higher strength increments were determined.

In a papaya plantation on Maui, where high and damaging populations of reniform nematodes had caused a reduction in fruit yield and quality, molasses applications lowered nematode soil populations and resulted in marked improvement in tree growth and harvestable fruit (Schenck, 2001). When applied to Chinese cabbage, there was a decrease in the numbers of *Heterodera* nematode cysts following harvest. Preplant applications of molasses to onions improved plant color and onion yield although no difference in soil nematode populations or in cyst number was observed. Molasses soil amendments supply carbohydrates and alter the C/N ratio. This affects the soil microbial ecology, usually resulting in lowered populations of plant parasitic nematodes as well as having other favorable effects on plant growth (Schenck, 2001). The specific mechanisms involved are not well understood and vary with the crops, soil conditions, and nematode species present.

Schaffert and Percival (2016) in the study of the Influence of biochar, slow-release molasses, and an organic N:P:K fertilizer on transplant survival of *Pyrus communis* 'Williams' *bon chrétien* indicated that, the use of biochar, slow-release molasses, and organic N:P:K fertilizer amendments offer potential for increasing bare-root transplant survival and establishment of *Pyrus communis* 'Williams' *Bon Chrétien.*

4.2. Bioethanol Production

Ethanol production from molasses is not new. Bioethanol has the same chemical make-up and characteristics as ethanol that has been produced by a chemical reaction e.g., Hydrogenation of ethane. Bioethanol is therefore suitable for all uses currently employed by ethanol. It can be used as a solvent, for which there is huge demand. Ethanol has many industrial uses due to its relatively high affinity both for water and a great range of organic compounds. It can be used in any extraction processes which require either a water or organic-solvent. It may also be added to solvent-based paints, lacquers, inks, household cleaning products, and, external pharmaceuticals (rubbing alcohol). The cosmetics industry uses alcohols in many different products. The most often used alcohol in cosmetics is ethanol. It is used in products such as toners and astringents, because of its solvent and antibacterial properties. One of the main uses that have recently been highlighted is that of Biofuels. It is non-petroleum-based alternative source of energy. Molasses fermentation-derived bioethanol is very suitable for use as a biofuel. Biofuels reduce green-house gases and produce less harmful emissions. Biofuels can be used in addition to, or as an alternative to petrol. It is suitable for use in petrol engines without the need for any alterations.

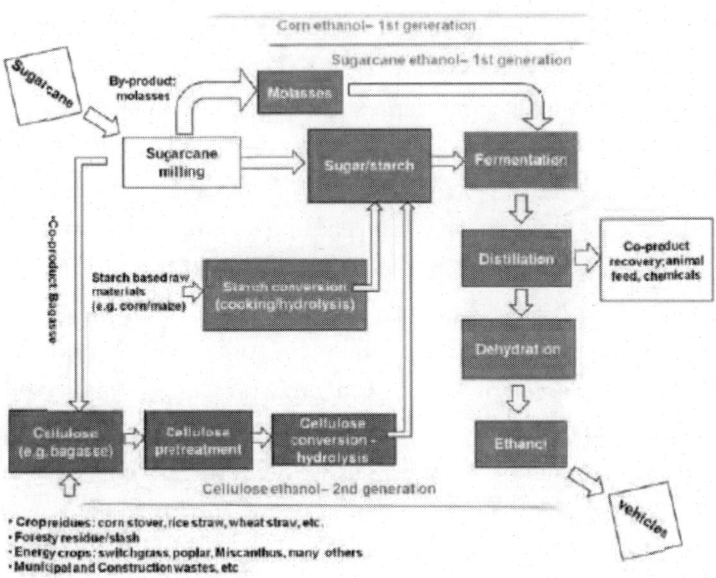

Source: Khatiwada (2010).

Figure 1. Production routes of bioethanol.

Bioethanol contributes more than 90% of the total liquid biofuel consumption in the world (IEA, 2007). Figure 1 shows different routes for bioethanol production process. As can be seen in the figure, sugar/starch obtained from sugarcane/corn or cellulose feedstocks follows the process of fermentation, distillation, and dehydration in order to produce first and second generation ethanol respectively. It should be noted that molasses (also a sugar based product) is treated separately in the figure in order to make it distinct from other conversion routes. This thesis covers molasses-based bioethanol. First generation bioethanol, obtained from sugarcane and corn feedstocks has so far dominated the bioethanol market globally and other technologies are at an early stage of development. In 2009, global ethanol production was about 74 billion liters, a four-fold increase since 2000 and it has contributed to a reduction in GHG emissions by 87.6 million tonnes in a year (RFA, 2010). The United States (US) was the world largest producer of bioethanol, accounting for about 52% (i.e., 38.5 billion liters) of the total bioethanol production in 2009. Brazil was the largest bioethanol exporter and second largest producer with a share of 34% (25 billion liters).

Ethanol from sugarcane has proven to be most cost competitive over the last few decades, following a steep learning curve in the production of ethanol, and there is a strong global ethanol market for international development (Hira, 2010; Khatiwada, 2010).

Bakhiet and Mahmoud (2015) in their study on the production of bioethanol from molasses by *Schizosaccharomyces* species showed that the moisture content of molasses was found to be 65%. The ash was 6.50%. The pH value was decreased by one unit during the fermentation processes due to the molasses degradation with acid production. Bio-ethanol was produced from two types of molasses preparations (raw molasses and sucrose determined concentration samples). *Schizosaccharomyces* spp. fermented molasses samples at all concentrations except, 100% because the solution was hypertonic and the microorganisms did not tolerate that concentration. The highest volume of ethanol obtained at concentration of 3:300 ml of molasses/row. While the lowest one obtained at concentration of 4:10% of sucrose/row. The final bioethanol appeared to be colourless, clear, bright, and free from turbidity indicating its high specification quality. They concluded that, the best conditions to obtain a highest volume of bioethanol are appropriate concentrations of molasses and suitable pH. The highest volume of bioethanol was 23.51 ml, obtained at 85.5 g/solids (molasses) and pH 6. While the highest volume of bioethanol is sucrose determined concentration sample was 16.03 ml at 71.25 g/solids and pH 6. We recommended the utilization of *Schizosaccharomyces* species in large scale production of ethanol to manage the industrial wastes.

4.3. Pharmaceutical Industry

Molasses is used in the pharmaceutical industry for the production of citric acid and lactic acid, among others. Citric acid ($CH_2COOH.COH.COOH.CH_2COOH$) is a tricarboxylic acid, soluble in water with a pleasant taste; it is an important acid used in food Industries. It exists in nature when carbohydrates are oxidized to carbon dioxide. Because of its high solubility, palatability and low toxicity it can be used in food, biochemical, and pharmaceutical industries. Bakhiet and Al-Mokhtar (2015) in their studies on citric acid production from fungi (*Aspergillus niger*) using by-product of sugar (sugarcane molasses) showed that citric acid production from the soil sample was of high amount, when compared with air, and bread. The soil sample produced 9.6% of citric acid compared with air 6.7% and bread 7.7%. The maximum citric acid production was produced on the 6th day of fermentation in all samples. By recycling and reusing waste material from cane molasses citric acid production can be easily achieved by using microorganisms that have the ability to produce citric acid efficiency such as *Aspergillus niger*.

Lactic acid can be used as a preservative, acidulant, and flavor in food, textile, and pharmaceutical industries. It could become a chemical commodity for the production of lactate esters, propylene glycol, propylene oxide, acrylic acid, 2,3-pentanedione, propanoic acidacetaldehyde, and dilactide (Vardarajan and Miller, 1999; Akerberg and Zacchi, 2000). Dumbrepatilet al. (2008) in their studies on the utilization of molasses sugar for lactic acid production by *Lactobacillus delbrueckii* subsp. *Delbrueckii* Mutant Uc-3 in batch fermentation concluded that the *Lactobacillus delbrueckii* mutant Uc-3 proved to be a promising strain for the production of lactic acid from molasses. The requirement for yeast extract is brought down to a minimum because molasses contains enough of the nitrogen source necessary to grow such fastidious organisms. Also, the buffering capacity of molasses could be an advantage in maintaining the pH of the medium above pH 5.3 during fermentation. Molasses was also proven to be an economically feasible raw material for industrial production of lactic acid since it is fortified with enough nutrients necessary for growth of lactic acid bacteria.

CONCLUSION

The added benefits of feeding molasses have been quantified by numerous research studies. There is no doubt that molasses is an excellent source of energy and minerals for ruminants. It can be fed in various ways and is very useful in many situations. Cattle and small ruminant producers can feel confident feeding molasses, knowing that they are feeding a safe and economical supplement. The added benefits of feeding molasses have been quantified by numerous research studies. There is no doubt that molasses is an excellent source of energy and minerals for ruminants. It can be fed in various ways and is very useful in many situations. Cattle and small ruminant producers can feel confident feeding molasses, knowing that they are feeding a safe and economical supplement. Thus, molasses, which is an excellent source of rapidly fermentescible energy, optimizes the use of ammonia produced by urea, and generates minerals (especially trace elements) will be a good supplement. Despite these positive effects of molasses on the use of poor forages, many studies remain to be carried out on the effect of molasses level in animal feeding. The essence of this paper is to contribute to the existing knowledge which is common and cheaper on how to improve the utilization of molasses by ruminants and non-ruminant in Cameroon to increase livestock production.

REFERENCES

Akerberg, C., and G. Zacchi. 2000. An economic evaluation of the fermentative production of lactic acid from wheat floor. *Bioresour.Technol.* **75:** 119–126.

Archimède, H., Garcia G., Xande X., Gourdine J. L., Renaudeau D., Despois E., Anais C., Coppry O., Fleury J., Mahieu M., Boval M., Alexandre G., Arquet R. et Gonzales-Garcia E. 2008. *Guide d'utilisation de la canne Guide d'utilisation de la canne à sucre et de sescoproduits à sucre et de sescoproduitsen alimentation animaleen alimentation animale: A l'usage des producteursagricoles et techniciens.* [Cane User's Guide Guide to Using Sugar Cane and Sugar Products and Its Feed Products in Animal Feed: For Farm Producers and Technicians]. INRA Institut National de la recherche Agronomique 78p.

Baker, F.S. 1954. Steer fattening trials in north Florida. *Florida Agr. Exp. Sta.* NFES Mimeo Rpt. 55-1.

Baker, F.S. 1955a. Citrus molasses, dried citrus pulp, citrus meal and blackstrap molasses in steer fattening rations. *Florida Agr. Exp. Sta.NFES Mimeo Rpt.* 55-3.

Baker, F.S. 1955b. Citrus molasses, dried citrus pulp, citrus meal, blackstrap molasses and stilbestrol in steer fattening rations. *Florida Agr. Exp. Sta. NFES Mimeo Rpt.* 56-1.

Baker, F.S. 1966. Ground snapped versus ground shelled corn with different roughages and with citrus pulp in steer fattening rations. *Florida Agr. Exp. Sta. NFES Mimeo Rpt.* 66-5.

Bakhiet Shami E. A. and Ehood A. I. Al-Mokhtar. 2015. Production of Citric Acid by *Aspergillus niger* Using Sugarcane Molasses as Substrate. *Jordan Journal of Biological Sciences.* 8(3):211-215.

Bakhiet, Shami, E. A., and Mahmoud Marwa Abdalrhim. 2015. Production of Bio-ethanol from Molasses by *Schizosaccharomyces* Species. *Annual Research & Review in Biology,* 7(1): 45-53, 2015.

Balch, C.C., and R.C. Campling. 1962. Regulation of voluntary food intake in ruminants. *Nutr. Abstr. Rev.* 32: 669-686.

Beardsley, D.W., F.M. Pate, B.W. Hayes, A.Z. Palmer and J.W. Carpenter. 1971. Raw sugar as an energy Source in steer finishing rations. *J. Anim. Sci.* 33: 706-710.

Bernard, M., Chapoutot, P., Chatelet, M., Gueroult, M., Jubert, M., Morel d'Arleux, F., Taccart, M., Mariani, M., Tierny, M. 1991. *Synthèse sur la mélasse: Comité des sous-produits-RNED bovins.* [Synthesis on Molasses: By-Products-RNED Cattle Committee]. Fiche N°8- Mélasse de betterave et de cane, Juillet. 19 pages.

Bisschoff, W.V.A., L.R. Quinn, G.O. Mott, and G.L. da Rocha. 1967. Supplemental feeding of steers on pasture with protein-energy supplements. *IRI Res. Instit. Bull.* 35.

Bohman, V.R., G.W. Trimberger, J.K. Loosli and K.L. Turk. 1954. The utilization of molasses in the rations of growing dairy cattle. *J. Dairy Sci.* 37: 284-293.

Bond, J., and T.S. Rumsey. 1973. Liquid molasses-urea or biuret (NPN) feed supplements for beef cattle: wintering performance, ruminal differences and feeding patterns. *J. Anim. Sci.*37: 593-598.

Bórquez-Gastelúm, J. L., Trujillo-Gutiérrez, D., Domínguez-Vara, I. A., Pinos-Rodríguez, J. M., Cobos-Peralta Mario, A. 2018. Yield performance of growing lambs fed silages with poultry litter, pig excreta and urea with molasses cane or a bakery by-product. *Agrociencia* 52: NÚMERO 3. 333-346. 2018.

Boukila, B., Tendonkeng, F.,Pamo, T. E., Matumuini, N.E. F., Mboko, A. V., Miegoue, E., et Allogo, A. R. 2013. Digestibilité *in vivo* des feuilles fraîches de bananier (*Musa sp.*) traitées à la mélasse chez les brebis djallonké (*Ovisaries*) [In vivo digestibility of fresh banana leaves (Musa sp.) Treated with molasses in djallonke ewes (Ovis aries)]. *Bull. Anim. Hlth. Prod. Afr.*, (2013),**61(1),** 101 – 107.

Bradley, N.W., J.R. Overfield and C.O. Little. 1966. Source of nitrogen for supplementing ground ear corn rations. *Kentucky Anim. Sci. Res. Rpt.* pp 16-18. *Kentucky Agr. Exp. Sta. Prog. Rpt.* 164.

Brannon, W.F., J.T. Reid, and J.I. Miller. 1954. The influence of certain factors upon the digestibility and intake of pasture herbage by beef steers. *J. Anim. Sci.* 13: 535-542.

Bray, C.I., M.G. Snell, F.L Morrison and M.E. Jackson. 1945. Feeding blackstrap molasses to fattening steers. *Louisiana Agr. Exp. Sta. Bull.* 394p.

Brown, G.H., J.R. Overfield, N.W. Bradley and C.O. Little. 1967. Source of nitrogen for supplementing ground ear corn rations for feedlot cattle. *Kentucky Anim. Sci. Res. Rpt.* pp 8-10. *Kentucky Agr. Exp. Sta. Prog. Rpt.* 190.

Brown, P.B. 1962. Sugar cane bagasse-blackstrap molasses rations for beef cattle. pp1216-1224. *Proc., 11th Congo Into Soc. Sugar Cane Techno.*

Brown, P.B. 1967. Roughage-molasses rations for wintering beef cattle. *Louisiana Agr. Exp. Sta. Bull.* 617.

Campbell, C.M., E.H. Cobb, O. Wayman and J. Lee. 1970. Effects of molasses and urea levels in feedlot rations for steers. *Proc., Western Sect. Am. Soc. Anim. Sci.* 21: 201-206.

Carlo, I., R. Mendoza and J. Velez. 1972. Evaluation of different feeding systems in beef cattle production. *J. Agr. Univ. Puerto Rico*. 46- 327-336.

Chapman, H.L. and J.E. Pace. 1974. Liquid feed mixtures for beef cattle. *Florida Agr. Exp. Sta. Circ.* S-229.

Chapman, H.L., R.W. Kidder, M. Koger, J.R. Crockett and W.K. McPherson. 1965. Blackstrap molasses for beef cows. *Florida Agr. Exp. Sta. Bull.* 701.

Chenost, M, Kayouli, C. 1997. *Utilisation des fourrages grossiers en régions chaudes* [Use of roughage in warm regions]. FAO. 226p.

Conrad, H.R. 1966. Symposium on factors influencing the voluntary intake of herbage by ruminants: physiological and physical factors limiting feed intake. *J. Anim. Sci,* 25: 227-235.

Conrad, H.R., A.D. Pratt and J.W. Hibbs. 1964. Regulation of feed intake in dairy cows. I. Change in importance of physical and physiological factors with increasing digestibility. *J. Dairy Sci.* 47: 54-62.

Cooper, D.P., R.D. Goodrich and J.C. Meiske. 1978. Comparison of hemicellulose extract with cane molasses in finishing beef rations. 1978 Minnesota Cattle Feeders' Rpt., *Minnesota Agr. Exp. Sta. Res. Rpt.* B-246.

Copeman, D.B., F.P. Sumbung, P. Poothongtong, S. Hag and J.R. Armstrong. 1977. Effect of molasses supplementation and copper, cobalt and anthelmintic therapy on weight gain of beef yearling steers grazing improved pastures in a wet tropical environment. *Australian J. Exp. Agr. Anim. Husb.* 17: 538-544.

Crawford, D.F., W.B. Anthony and R.R. Harris. 1978. Evaluation of concentrated hemicellulose extract as cattle feed. *J. Anim. Sci*. 46: 32-40.

Creek, M.J., D.M. Redfern and H.A. Squire. 1974. Intensive cattle feeding in Kenya using high levels of molasses. *World Rev. Anim. Prod.* 10(4): 58-68.

Curtin, L.V. 1983. Molasses-general considerations Excerpted from *Molasses in Animal Nutrition*. National Feed Ingredient Association West Des Moines, Iowa.

Davis, R.F., G.W. Timberger, K.L. Turk and J.K. Loosli. 1955. Feeding value and digestibility of cane molasses nutrients for dairy heifers. *Cornell Agr. Exp. Sta. Bull.* 914.

Delgado, A., A. Elias, J.L. Veitia and F. Alfonso. 1975. The use of pasture for beef production. 3. Different protein sources for molasses/urea supplementation to bulls during the dry season. *Cuban J. Agr. Sci*. 9: 255-262.

Delgado, A., J.L. Veitia, A. Elias and R. Dieguez. 1978. The use of pasture for beef production. 5. Supplementation with molasses, urea and true protein for yearling calves fed with low quality forage. *Cuban J. Agr. Sci.* 12: 143-150.

Dumbrepatil Arti, Mukund Adsul, Shivani Chaudhari, Jayant Khire, and Digambar Gokhale. 2008. Utilization of Molasses Sugar for Lactic Acid

Production by *Lactobacillus delbrueckii* subsp. *Delbrueckii* Mutant Uc-3 in Batch Fermentation. *Applied and Environmental Microbiology*, 74 (1):333–335. doi:10.1128/AEM.01595-07.

Elias, A., T.R. Preston and M.B. Willis. 1969. Intensive beef production from sugar cane. 8. The effect of rumen inoculation and different levels of forage on the performance of Brahman bulls fattened on high levels of molasses/urea. *Rev. Cubana Cienc. Agr.* (Eng. Ed.) 3: 19-23.

Findlay, M. Pate, 2003. Molasses in beef nutrition Article Was Excerpted From *Molasses in animal nutrition* Copyright 1983 National Feed Ingredients Association West Des Moines, Iowa.

Fogang Zogang, B., Benoît Boukila, Camara Sawa, Fernand Tendonkeng, Zougou Tovignon G., and Etienne Pamo Tedonkeng. 2012. Physical characteristics and palatability of multinutritional Tithonia diversifolia blocks associated with Brachiaria ruziziensis straw in the Djallonke ewe. *Livestock Research for Rural Development. Volume 24, Article #088.* Retrieved, from http://www.lrrd.org/lrrd24/3/foga11003.htm.

Fogang Zogang, B., Tendonkeng, F., Camara Sawa, Boukila, B., Pamo, T. E. 2013. Effects of inclusion of different levels of Tithonia diversifolia leaf in multinutritional blocks on the intake and in vivo digestibility of Brachiaria ruziziensis straw rations in the Djallonke ewe. *Revue d'élevage et de médecine vétérinaire des pays tropicaux*, 2013, **66 (3):** 103-107.

Gaili, E.S.E. and M.A. Ahmed. 1980. Feedlot performance, carcass and offal yield of Kenana x Friesian bulls finished on rations containing different levels of molasses. *Trop. Anim. Prod.* 5: 248-252.

Hart, R.H., J. Bond, G.E. Carlson and T.R. Rumsey. 1971. Feeding corn or molasses to cattle on orchard grass pasture. *Agro. J.* 63: 397-401.

Hatch, C.F. and W.M. Beeson. 1972. Effect of different levels of cane molasses on nitrogen and energy utilization in urea rations for steers. *J. Anim. Sci.* 35: 854-858.

Heinemann, W.W. and E.M. Hanks. 1977. Cane molasses in cattle finishing rations. *J. Anim. Sci.* 45: 13-17.

Hira, A., 2010. Sugar rush: Prospects for a global ethanol market. *Energy Policy.* doi:10.1016/j.enpol.2010.05.035.

Holder, N.L. 1972. *Supplemental feeding of beef cattle grazing pangola grass pastures on the intermediate savannahs of Guyana*. M.Sc. Thesis. University of Florida, Gainesville.

IEA, 2007. IEA Energy Technology Essentials - Biofuel Production. *International Energy Agency* (IEA). Available at http://www.iea.org/techno/essentials2.pdf.

James, L.A. 1973. Comfith in rations for livestock. *Can. Internat'l. Devel. Agency Seminar on Sugarcane as Livestock Feed, Barbados*, Jan. 30-31, 1973.

Kargaard, J. and B.D.H. Van Niekerk. 1977. Incorporation of DPW, urea and fish meal with varying molasses levels in cattle feedlot rations. *S. Afri. J. Anim. Sci.* 7: 117-125.

Khatiwada Dilip, 2010. *Assessing the sustainability of bioethanol production in Nepal.* Licentiate Thesis, Department of Energy and Technology, KTH School of Industrial Engineering and Management. Stockholm, Sweden. 79p.

King, W.A., C.C. Brannon and H.J. Webb. 1960. Blackstrap molasses for dairy heifers fed grass hay or silage. *South Carolina Agr. Exp. Sta. Bull.* 479.

Kirk, W.G., E.M. Kelly, H.J. Fulford and H.E. Henderson. 1966. Feeding value of citrus and blackstrap molasses for fattening cattle. *Florida Agr. Exp. Sta. Bull.* 575-A.

Lemoufouet, J., Tendonkeng, F., Miegoue, E., Azoutane, J., Matumuini, N.E.F., Fogang, Z.B., Mboko, A.V., Boukila, B., Pamo, T.E. 2014a. Effect of molasses on ingestion and digestibility of maize stover treated with 28% chicken manure in small ruminants. *Int. J. Biol. Chem. Sci.*, **8(4):**1581-1593.

Lemoufouet, J., Tendonkeng, F., Miégoué, E., Soumo, S.N., Mbainaissem, B., Fogang Zogang, B., Mboko, A.V., Matumuini, F.N.E., Boukila, B., and Pamo, E.T.2014b. Ingestion and digestibility in sheep of urea-treated maize stover associated with molasses. *Livestock Research for Rural Development. Volume 26, Article #45.* Retrieved February 27, 2014, from http://www.lrrd.org/lrrd26/3/lemo26045.html.

Lemoufouet, J., Tendonkeng, F., Miégoué, E., Fogang, Z.B., Matumuini, N.E.F., Mboko, A.V., Boukila, B., and Pamo, T.E. 2016. In vitro digestibility of maize stover treated with urea or chicken droppings associated with molasses in small ruminants. *Livestock Research for Rural Development. Volume 28, Article #27.* Retrieved February 2, 2016, from http://www.lrrd.org/lrrd28/2/lemo28027.htm.

Lhoste, Ph., 1980. L'utilisation de la mélasse du Cameroun enembouche bovine - expérimentation sur différents types d'animauxbœufs, vaches de réforme et taurillons ruminants [The use of Cameroon's beef molasses - experimentation on different types of cattle, cull cows and ruminant young bulls]. I.E.M.V.T. PP 450-459.

Lishman, A.W. 1967. Cane molasses as a substitute for maize in beef finishing rations. *S. Afri. J. Agr. Sci.* 10: 51-60.

Lofgreen, G.P. 1965. Net energy of fat and molasses for beef heifers with observations on the method for net energy determination. *J. Anim. Sci.* 24: 480-487.

Lofgreen, G.P. and K.K. Otagaki. 1960a. The net energy of blackstrap molasses for fattening steers as determined by a comparative slaughter technique. *J. Anim. Sci.* 19: 392-403.

Martin, J.L., T.R. Preston and M.B. Willis. 1968. Intensive beef production from sugar cane. 6. Napier or maize as forage sources at two levels in diets based on molasses/urea. *Rev. Cubana Cienc. Agr.* (Eng. Ed.) 2: 175-181.

Martin, P.C. and F. Alfonso. 1978. Effect of restricted molasses- urea supplementation on bulls grazing pangola grass (*Digitaria decumbens* Stent). *Cuban J. Agr. Sci.* 12: 51-58.

Matumuini, F.N.E., Tendonkeng, F., Mboko, A.V., Zougou, G.T., Miégoué, E., Boukila, B. and Pamo, T.E. 2013. Ingestion and in vivo digestibility of Tithonia diversifolia leaves treated with molasses associated with maize stover in the Djallonke ewe (Ovis aries). *Livestock Research for Rural Development. Volume 25, Article #142.* Retrieved, from http://www.lrrd.org/lrrd25/8/matu25142.htm.

Matumuini, F.N.E., Tendonkeng, F., Mboko, A.V., Zougou, G.T., Miégoué, E., Lemoufouet, J., Boukila, B. and Pamo, T.E. 2014a. In vitro digestibility of maize stover associated with Mexican sunflower (Tithonia diversifolia) leaves treated with molasses in small ruminants. *Livestock Research for Rural Development. Volume 26, Article #106.* Retrieved June 2, 2014, from http://www.lrrd.org/lrrd26/6/matu26106.html.

Matumuini, N. E. F., Tendonkeng, F., Mboko, V. A., Zougou, T. G., Miegoue, E., Lemoufouet, J., Ndongo, D. T., Boukila, B., Pamo, T. E. 2014b. Effect of Tithonia diversifolia leaf treatment on molasses on the ingestion and digestibility of maize stover in West African dwarf goat (Capra hircus hircus). *Int. J. Biol. Chem. Sci.,* 8(1): 249-258.

Merrill, W.G., S. Lovejoy, K.L. Turk, K.E. Harshbarger, J.K. Loosli and G.W. Trimberger. 1959. Value of cane molasses and urea fed with early- and late-cut hay in a ration for dairy heifers. *Cornell Agr. Exp. Sta. Bull.* 937.

Molina, A. 1977. Feedlot performance and carcass traits of Santa Gertrudis and Charolais x Brahman bulls fattened with green forage, final molasses or concentrates. *Cuban J. Agr. Sci.* 11: 277-287.

Morciego, S., F. Munoz and T.R. Preston. 1970. Commercial fattening of bulls with molasses/urea and restricted grazing. *Rev. Cubana Cienc. Agr.* (Eng. Ed.) 4: 97-100.

Mott, G.O., L.R. Quinn, W.V.A. Bisschoff and G.L. da Rocha. 1967. Molasses as an energy supplement for zebu steers grazing nitrogen-fertilized and unfertilized coloniaquineagrass pasture. *IRI Res. Instit. Bull.* 36.

Munoz, F., S. Morciego and T.R. Preston. 1970. Commercial fattening of bulls on molasses/urea, fish meal and restricted forage under feedlot conditions. *Rev. Cubana Cienc. Agr.* (Eng. Ed.) 4: 91-96.

Olbrich, S.E. and 0. Wayman. 1972. Effect of feeding raw sugar on growth performance and rumen fluid parameters of fattening beef cattle. *J. Anim. Sci*. 34: 820-825.

O'Mary, C.C., A.E. Cullison and J.L. Carmon. 1959. Implanted and oral stilbestrol for fattening steers. *J. Anim. Sci*. 18: 14-20.

Porres, G. 1971. Molasses and molasses/urea as supplements for F1 Holstein x Brahman bulls grazing highly fertilized pangola (Digitariadecumbens) pasture. *Rev. Cubana Cienc. Agr.* (Eng. Ed.) 5: 13-20.

Potter, G.D., C.O. Little, N.W. Bradley and G.E. Mitchell. 1971. Abomasal nitrogen in steers fed soybean meal, urea or urea plus two levels of molasses. *J. Anim. Sci.* 32: 531-533.

Preston, T.R. 1969. Symposium on beef from the tropics: 3. Beef from sugar cane. *Rev. Cubana Cienc. Agr.* (Eng. Ed.) 3: 141-153.

Preston, T.R. and M.B. Willis. 1970a. A new look at molasses for livestock feeding. *Feedstuffs* 42 (13): 20-25.

Preston, T.R., M.B. Willis and A. Elias. 1967b. Intensive beef production from sugar cane. 2. Comparison between normal and invert molasses as a supplement to forage or concentrate. *Rev. Cubana Cienc. Agr.* (Eng. Ed.) 1: 41-48.

Preston, T.R., M.B. Willis and A. Elias. 1970b. The performance of two breeds given different amounts and sources of protein in a high-molasses diet. *J. Anim. Prod.* 12: 457-464.

Preston, T.R., M.R. Willis and J.L. Martin. 1969. Efficiency of utilization for fattening of the metabolizable energy of molasses-based diets. *J. Anim. Sci.* 28: 796-801.

Ravi E., Animesh Sharma, A.T. Manikandan, G. karthick, A. Abdul Jameel. 2015. Study on Effect of Molasses on Strength of Soil. *International Journal of Advanced Research Trends in Engineering and Technology (IJARTET) Vol. 2, Issue 2, February 2015.*

RFA, 2010. Ethanol industry outlook - climate of opportunity. *Renewable Fuels Association* (RFA). Available at http://www.ethanolrfa.org/page/-/objects/pdf/outlook/RFAoutlook2010_fin.pdf?nocdn=1.

Riggs, J.K. and A.Y. Blankenship. 1955. Comparative value of five types of molasses for fattening beef cattle. *Texas Agr. Exp. Sta. Prog. Rpt.* 1820.

Rush, I.G. and R. Totusek. 1976. Supplemental value of feed grade biuret and urea-molasses for cows on dry winter grass. *J. Anim. Sci.* 42: 497-505.

Wayman, O., I.I. Iwanaga, K. Morita and L.A. Henke. 1952. Sugar cane by-products as products as major constituents of rations for dairy heifers, milking cows, and beef steers. *Hawaii Agr. Exp. Sta. Prog.Note* 84.

Wayman, O., I.I. Iwanaga, L.A. Henke and H.J. Weeth. 1953. Fattening steers on sugar cane by-products. *Hawaii Agr. Exp. Sta. Circ.* 43.

Wayman, O., L.A. Henke and I.I. Iwanaga. 1954. Further studies on the use of cane molasses rations for dry-lot finishing of beef cattle. *Hawaii Agr. Exp. Sta. Prog. Note* 103.

Zelakwa, M. 2013. *Rapport sur les agro-industries dans le développement des zones rurales et dans les stratégies de développement au Cameroun: cas de la SOSUCAM* [*Report on agro-industries in rural development and development strategies in Cameroon: case of SOSUCAM*]. CIRAD, 2013. 36p.

In: Molasses: Forms, Production and Uses ISBN: 978-1-53614-703-2
Editors: K. Maddison and R. Fuller

Chapter 5

THE POTENTIAL OF MOLASSES TO ADD VALUE IN FOOD PROCESSING

Bojana Filipčev*[*], *PhD
Institute of Food Technology, University of Novi Sad, Novi Sad, Serbia

ABSTRACT

Molasses is a by-product of industrial processing of sugar from sugarcane or sugar beet. The sugar processing generally consists of serial steps of repeated evaporation, crystallization and centrifugation of cane or beet extraction juices. Molasses remains as dark, viscous syrup following the phases of crystallization and separation of raw sugar. Molasses is a polycomponent system of variable composition due to many factors: biological origin (cane or beet), raw material quality, applied processing methods during juice clarification and sugar crystallization, etc. The main constituents of molasses are fermentable sugars (saccharose, glucose, and fructose) which proportion depends on the nature of molasses (beet or cane). The non-sugar part of molasses is abundant in minerals, especially potassium, sodium, magnesium, calcium and iron, and contains a myriad of versatile other bioactive compounds such as B group vitamins, choline, allantoin, purine, cytosine, guanosine, cytidine, glutamine acid, lactic acid,

[*] Corresponding Author Email: bojana.filipcev@fins.uns.ac.rs

pectin, phenolics, etc. Beet molasses contains betaine which has been recently recognized as a functional compound with proven beneficial health effect. A few studies highlighted molasses as a health promoting ingredient due to potent antioxidative activity and proposed its use as a valuable nutritious, yet affordable, commodity.

Unlike beet molasses, cane molasses has been used, though not extensively, in bakery and confectionery industry, mainly as a minor ingredient to provide sweetness, unique flavor characteristics and color to a product. Molasses may afford some other valuable functions such as shelf-life improvement, enhancement of leavening activity and buffering capacity. Beet molasses currently has no use in human consumption, mainly due to strong, earthy flavor which is absent in cane molasses. Although odd flavor of beet molasses disables its consumption on its own, several studies have indicated successful use of beet molasses in meat and bakery products. The studies demonstrated the applicable molasses doses and their effects on the various aspects of product quality. It was confirmed that beet molasses can be used in various food products without having adverse effects on product palatability but providing an outstanding improvement in nutrient pattern and antioxidative potential.

The aim of the present chapter is to review the latest advances on the potential of molasses (beet or cane) as a source of functional ingredients as well as its application in various food products (bakery, meat, vegetables and fruit products).

Keywords: sugar beet, sugarcane, composition, antioxidants, minerals, phenolics, functional food, health effects

1. INTRODUCTION

In recent times, there has been an increased interest for recovery of valuable bioactive compounds from waste materials generated during food processing. Food by-products are interesting materials due to low price, abundant quantities and accumulation at few locations (Galanakis, 2012). Further processing or re-use of industrial waste is useful from an environmental point of view, in addition to its revalorization in the form of novel value-added products.

Molasses is a waste stream generated in the process of sugar refining either from sugarcane or sugar beet. In spite of differences in composition and processing stages, both beet and cane molasses have been recognized as rich sources of potent bioactive compounds: macroelements, microelements, antioxidants, organic acids, biotin, B group vitamins, phenolics, flavonoids, purine and pirimidine bases, etc. Scientific research has been confirming biological functionality of molasses or molasses extracts. Many effects have been reported: antioxidant, antimutation, anti-inflammatory, tyrosinase inhibitory, vaccine adjuvant, immunomodulatory, antitumor, and infection resistant effects. It has been even suggested that molasses is superior to honey regarding nutritional and functional properties (Šušić & Sinobad, 1989).

Current use of molasses has been mainly limited to its use as a livestock feed, soil fertilizer, and a substrate for fermentation industry to produce ethanol, baker's and brewer's yeast, citric acid, lysine and monosodium glutamate. Unlike beet molasses, cane molasses has been used in culinary practice and food industry (bakery, confectionery) as a minor ingredient, mainly as a taste, sweetness and color enhancer. However, molasses can contribute to other functionalities in the product and this makes it interesting for further research.

2. Production and Classification of Beet and Cane Molasses

Molasses originates from the process of sugar refining from two different raw materials, sugar beet or sugar cane, therefore the basic classification of molasses into cane or beet molasses. Eighty percent of world's sugar production comes from sugarcane while the rest 20% accounts for sugar beet. In Europe, the dominating raw material for sugar production is sugar beet whereas sugar cane prevails in Cuba, Brazil, Mexico, Australia, India, South Africa, Thailand and Indonesia. The process of sugar refining is essentially similar for both crops though some differences exist. In brief, after the harvest of the crop, it is sliced (beet) or crushed (cane) and subjected to extraction with water to obtain extraction juice. Extraction juices are cleaned to remove impurities and then boiled to concentrate and enhance sugar crystallization. Sugar crystals are removed from liquid syrup by centrifugation. Simplified schemes of the sugar refining process from beet and cane are displayed in Figures 1 and 2. The major difference between the processes is related to the collection of molasses. Molasses from sugar beet is generated as the final remnant of the process from which it is not economical to further exhaust sugar. On the other hand, cane molasses can be collected after each extraction-crystallization cycle. Therefore, cane molasses is available in various grades and characteristics, unlike beet molasses.

2.1. Classification of Cane Molasses

Table 1 displays some of the classifications of cane molasses and terms used to denote cane molasses and similar products. As it will be seen from the explanations below, the great number and variability of terms comes from the fact that many products denoted as cane molasses are, actually, not true molasses or by-products of sugar refining but semi-products obtained by concentration of extraction juices (usually cane juice) either prior or after partial sugar removal.

The USDA Standard defines sugarcane molasses as “the clean, sound, liquid product obtained by evaporating the juice of sugarcane and the removal of all or any part of the commercially crystallizable sugar” (USDA, 1959) and distinguishes 4 grades of cane molasses (A, B, C and substandard). This classification refers to *final, blackstrap molasses* and defines feed-grade material (Clarke, 1993).

High-test molasses is produced from clarified or unclarified sugar cane juices (Perez, 1995). Strictly speaking, high-test molasses is not a by-product of sugar production but an inverted syrup since it is obtained by concentrating cane juice to around 85° Brix with subsequent inversion with either acid or invertase (Clarke, 1993; Olbrich, 1963). Due to high sugar concentration, less thermal degradation products and the presence of compounds native to cane, high-test molasses is a premium product with emphasized cane aroma (Clarke, 1993). Bruhns (1998) lists high-test molasses as edible molasses.

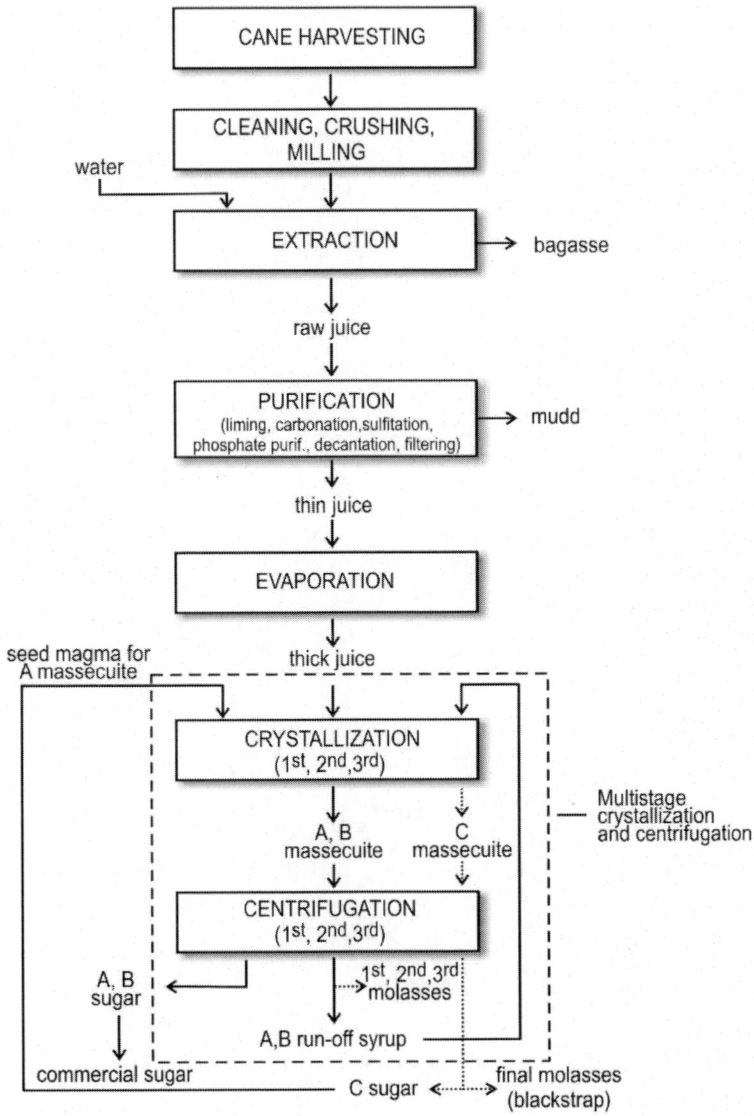

Figure 1. Simplified flow diagram of sugar production from sugarcane.

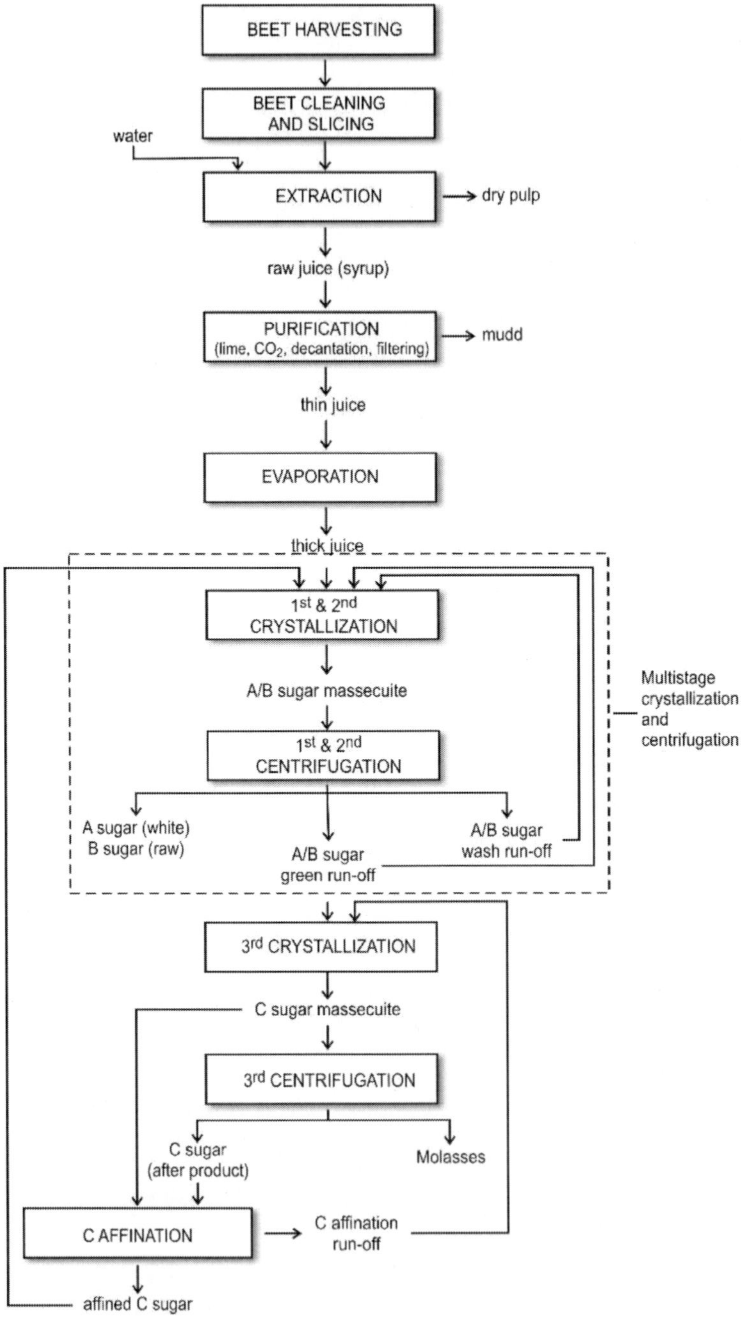

Figure 2. Simplified flow diagram of sugar production from sugar beet.

Table 1. Cane molasses types and terminology used to denote the types

Sugarcane molasses types	Reference
A Grade (U.S. Fancy) B Grade (U.S. Choice) C Grade (U.S. Standard) Substandard molasses	USDA (1959)
Unclarified (integral) high-test molasses High test molasses A, B, C molasses Blackstrap (treacle)	Perez (1995)
Imported unsulfured molasses Mill molasses: Whole juice molasses (WJM) First, second, third molasses Final molasses (blackstrap) Refiner's molasses Dry molasses	Hickenbottom (1996)
Edible molasses (high-test molasses) Refinery molasses Vinasse or condensed molasses	Madsen (1998)
Cane syrups, golden syrups Molasses: Blackstrap High-test molasses (Fancy molasses, cane invert syrup, cane juice molasses) Sulfured/unsulfured molasses Commercially available molasses	Clarke (1993)
Light molasses (original, mild or Barbados) Dark molasses (robust, full flavored, cooking molasses) Blackstrap molasses Sulfured and unsulfured molasses Fancy molasses Cooking molasses	Common culinary classification Haney (2018); Google (2018a)

A/B molasses or *1st/2nd molasses* are intermediate products obtained after centrifuging of massecuite in first and second steps of crystallization. They are actually run-off syrups which need partial inversion if to be stored (especially A molasses). *C molasses* is the final, blackstrap molasses or treacle (Perez, 1995).

Imported unsulfured molasses originating from Caribbean area is not a real molasses as it contains all the sugar present in raw syrup. It is prepared by clarifying and evaporating to 79.5% solids the raw cane syrup obtained by pressing. After controlled maturing and curing it may develop a rum-like flavor (Hickenbottom, 1996). It is the sweetest type of molasses, mellow in taste, mild in aroma, clear and light colored.

Mill molasses originates from cane producing states in the USA (Louisiana, Florida and Texas) and is further classified into 5 categories: whole juice molasses, first molasses, second molasses, third molasses, final (blackstrap) molasses (Hickenbottom, 1996). *Whole juice molasses* (WJM) is prepared from whole unextracted cane juice by clarification and evaporation to 79.5% solids and is practically similar to the imported unsulfured molasses. The difference between the two molasses is in flavor and color; WJM has harsher flavor and darker color due to climatic differences related to cane ripening between the two growing regions (Hickenbottom, 1996). The subsequent categories of molasses are obtained after extraction of sugar from WJM. Depending on the cycle of extraction, molasses can be *first, second, third* and *final*. *First molasses* is obtained from the run-off syrup after the first sugar extraction. In comparison to WHM, it has less delicate flavor and slightly darker color. *Second* and *third molasses* come from run-offs after succeeding sugar extractions and each one is darker and stronger in flavor. *Final molasses (blackstrap)* is the residual, maximally exhausted syrup in terms of sugar content. It is characterized with the darkest color and strongest bitterness (Hickenbottom, 1996).

Refiner's molasses is produced by evaporating the syrup extracted from raw sugar. It has no cane flavor but caramelized sugar flavor (Hickenbottom, 1996).

The *"sulfured"* and *"unsulfured"* labels denote whether sulfur dioxide was used in the juice purification stage or not (Madsen, 1998). The addition of SO_2 reduces the formation of color in thin juice. It also affects the taste of molasses by reducing its richness. Unsulfured molasses is considered of higher quality due to better taste and consistency. The production of SO_2-free sugar and consequently molasses requires pH stability of juice in alkaline region, absence of invert sugar and some adjustments in the boiling and evaporation stages (Madsen, 1998).

Dry molasses is a powdered form of molasses. Molasses is difficult to dry and drying agents such as dextrins, wheat flour or starch are added as carriers (Hickenbottom, 1996). Today, new forms of dry molasses are available on market such as granulated, free-flowing molasses which is made through co-crystallization process. Dry molasses granules are made from sugar and cane molasses and do not contain carriers or flow agents. Dry molasses is usually prepared from run-off products at higher extraction levels because higher invert sugars impair drying efficiency (Hickenbottom, 1996). Molasses is dried in drum or spray dryers (Hickenbottom, 1996). Dry molasses is hygroscopic and difficult to store, however, the use of adequate carriers makes molasses less hygroscopic and more resistant to caking. Its main advantage is that it is easier to handle in food applications as compared to sticky and viscous molasses. However, one should be aware that dry molasses is usually not 100% molasses due to presence of carriers. After rehydration, dry molasses results in a product with dark color and strong flavor as being made from streams at the lower end of extraction (Hickenbottom, 1996).

Culinary Definitions

Molasses intended for retail purchase or food industry use, cover a variety of products: light, dark, blackstrap, fancy, cooking molasses, etc. (Table 1). Most commercially available molasses is made by blending various cane factory or refinery molasses and syrups to obtain product with consistent and defined quality (Clarke, 1993). In culinary practice molasses is used to improve color, sweetness and flavor of foods. Molasses produced after the first boiling and extraction process is referred to as *light molasses*. This molasses type is the sweetest, light in color and texture, subtle in taste. Due to its mellow taste, it is a preferred ingredient in baking, marinades, rubs, sauces, oatmeal, and pancake toppings. *Dark molasses* is the result of the second boiling and extraction. It is less sweet, darker and more viscous. It is mainly used to flavor sweet baked goods like cookies, gingerbread, fruit cakes, pies, puddings but it is also suitable to improve the flavor of sauces, baked beans, meat, etc. (Google, 2018a). Depending on consumer's preferences, light and dark molasses can be used interchangeably. *Black-*

strap molasses and *treackle,* as the most concentrated form of cane molasses, dark and bitter in taste, is less frequently used to flavor food but it is sold in health food stores due to believed best health benefits. Blackstrap molasses is abundant in minerals, especially potassium, calcium, magnesium, iron and manganese and is a source of B6 vitamin (Google, 2018b). However, the widely believed health benefits of molasses have not been fully substantiated, though several studies infer that potent antioxidant activity of molasses may provide solid scientific basis to support the believed health claims. *Fancy molasses* is the sweetest type of molasses which is used as a topping for biscuits and pancakes (Haney, 2018). It is not a true molasses as it is not produced after sugar extraction but by concentration of pure cane juice, like WJM and high-test molasses. According to Clarke, fancy and high-test molasses are synonyms (Clarke, 1993). *Cooking molasses* is a blend of fancy and blackstrap molasses (Haney, 2018). It provides less sweetness and a full flavor.

3. Composition of Beet and Cane Molasses

3.1. Proximate Composition

The published data on proximate composition of beet molasses and different types of cane molasses is presented in Table 2. The characteristics and composition of molasses depend on many factors such as the biological origin (sugar beet or sugarcane, effect of variety), growing conditions (climate, harvest time), applied agricultural practices (irrigation, fertilization, storage conditions prior processing) as well as the differences in the processing, particularly in the stage of juice purification and massecuite crystallization (Higginbotham & McCarthy, 1998; Šušić & Sinobad, 1989). Therefore, molasses is a product with composition variable over wide ranges for which it is hard to ensure consistency.

Molasses is a complex mixture of sugars and nonsugars. The dominant constituent is sucrose, while other saccharides are present in lower amounts: glucose, fructose, raffinose (in beet molasses only), kestose, theanderose, starch, dextrans, levans, and inositol (mainly in beet molasses). The saccharose accounts to between 30%-60% (w/w) while glucose and fructose are present in amount ≤1% (w/w) in beet molasses and ≈15% (w/w) in cane molasses. Nonsugar substances in molasses include all compounds other than sugar which are maintained from the original crop, produced by chemical and enzymatic reactions during the processing and/or added throughout the processing. The nonsugar part of molasses can be classified into organic and inorganic substances (Olbrich, 1963). The nonsugars represent a versatile class of compounds and elements which constituents are listed in Figure 3. Beet molasses contains somewhat higher content of nonsugars (around 20% of total mass) whereas cane molasses contains 12-18% nonsugars (Higginbotham & McCarthy, 1998). Molasses does not contain fibers and fat. It has high content of solids, between 70 and 85% (Table 2). The ash content is high, up to 15% and reflects high mineral content (Table 2).

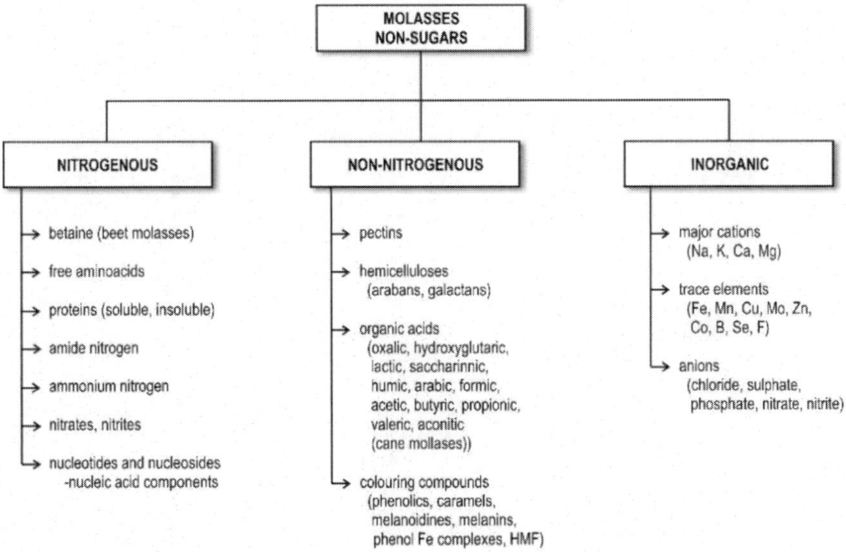

Figure 3. Constituents of molasses non-sugars (data collated from Higginbotham & McCarthy, 1998; Šušić & Sinobad, 1989; Olbrich, 1963).

Table 2. Composition of cane and beet molasses

Molasses type		Solids (%)	Total sugars (%)	Saccharose (%)	Invert sugar (%)	Proteins* (%)	Ash (%)	pH	Reference
Cane molasses and sugar products	blackstrap	79.5	53.0	34.0	19.0	2.2	9.5	5.0	Hickenbottom (1996)
	blackstrap	84.0 (°Bx)	49.0	32.0	17.0	-	12.7	5.8	Gasmalla, Yang, Nikoo, & Man (2012)
	blackstrap	75.0-83.0	-	30.0-40.0	10.0-25.0	2.5-4.5	7.0-15.0	-	Clarke (1993)
	A run-off	70	-	55.25	10.0	-	6.0-9.00	-	Higginbotham & McCarthy (1998)
	B run-off	70	-	45.50	13.0	-	9.0-11.0	-	Higginbotham & McCarthy (1998)
	C run-off	82	-	31.50	15.0-18.0	-	11.0-15.0	-	Higginbotham & McCarthy (1998)
	high-test molasses	81.0-92.0	86.0-79.0	-	-	0.31-1.56	1.8-3.6	6.0	Olbrich (1963)
	unsulfured molasses	-	72.0	35.0	37.0	-	2.5	-	Clarke (1993)
	all-purpose molasses	-	63.0-67.0	33.0-37.0	28.0-32.0	-	4.5-5.5	-	Clarke (1993)
	robust molasses	-	51.0-55.0	33.0-37.0	16.0-20.0	-	8.0-9.0	-	Clarke (1993)
Beet molasses		83.3	50.8	49.7	1.15	-	12.6	7.1	Šušić, et al. (1995)
		75.7	46.6	-	-	11.0	9.8	-	Sauvant, Perez, & Tran (2004)
		81.0-84.0	50.0-52.5	46.0-52	0.5	5.0-13.0	11.0-12.0	-	Šušić & Sinobad (1989)
		75.9-85.4	-	57.9-64.1	0.19-0.89	10.19-18.41		5.5-8.4	Olbrich (1963)

* (N x 6.25)

Important parameters to define the quality of molasses in trade are total sugars, invert sugars, solids (dry matter content), pH and volatile acids. Evaluation of these parameters is important as they affect the stability of molasses and may point to quality changes occurring during storage. Molasses in trade has to meet different quality specifications depending on its further use (as animal feedstock, in fermentation industry, for particle agglomeration, etc.). Higginbotham and McCarthy (1998) mentioned that slightly different standards exist regarding total sugar and solids content for molasses in trade throughout Europe and USA. The figures are 47-48% total sugars and 74-77% total solids for molasses in Europe and 48-50% total sugars and 80-84% solids in the USA. Feed-grade blackstrap molasses is defined by the USDA standard (Table 1). U.S. Grade molasses should comply with the requirements of the standard for solids (Brix), total sugars, ash, total sulfites, color and flavor. The standard describes the minimum requirements: for Grades A-B, solids are min 79° Brix, total sugars are in the range 63.5-58.0%, ash 5.25-10.00%, total sulfites 200-250 mg/kg, respectively (USDA, 1959). Substandard quality molasses is out of these ranges and has less solids and total sugars but more ash and sulfites (Clarke, 1993). In Serbia, minimal quality requirements for beet molasses in trade include 76.3° Brix solids, 46% sugars (polarimetrical), pH in range 7-8 and maximum sulfur dioxide content 70 mg/kg (Serbian Regulation, 2013; Serbian Standard, 1963).

3.2. Minerals and Vitamins

Minerals and vitamins are very important micronutrient components of molasses (Hickenbottom, 1996). Both beet and cane molasses contain appreciable amounts of minerals (8-15%) (Higginbotham & McCarthy, 1998). Mineral content is variable and mostly depends on the growing conditions of the original crop but employed processing methods can also affect the content of some minerals. Dominant minerals in molasses are potassium, sodium, calcium and magnesium. Generally, beet molasses is richer in potassium and sodium whereas cane molasses has more calcium and phosphorous. Primary macroelement in both cane and beet molasses is potassium. Potassium is an essential nutrient for some microorganisms and animals while in human nutrition its intake is globally well below recommendations in adult population. Potassium is not removed during sugar processing, unlike calcium and magnesium which are partially precipitated during juice purification. According to Šušić and Sinobad (1989), minerals in molasses are in bioavailable form which may offer considerable health benefits and makes molasses highly recommendable for human nutrition.

Table 3. Minerals in different molasses types

Type of molasses		K	Na	Ca	Mg	Fe	Zn	Mn	Reference
		mg/100 g				mg/kg			
Cane molasses and sugar products	"typical" molasses	1500-6000	20-400	300-900	50-1000	30-500	4-48	10-50	Higginbotham & McCarthy (1998)
	blackstrap	1380	192	517	-	75	-	-	Hickenbottom (1996)
	first molasses	875	25	220	-	15	-	-	Hickenbottom (1996)
	third molasses	1060	37	290	-	60	-	-	Hickenbottom (1996)
	refiner's molasses	950	25	225	-	25	-	-	Hickenbottom (1996)
	not specified	2400	200	800	-	24.9	-	-	Curtin (1983)
Beet molasses		2190-3820	710-1480	340-1020*	20-170*	-	-	-	Jevtić-Mučibabić, et al. (2013)
		3920	680	100	50	117	-	-	Sauvant, Perez, & Tran (2004)
		3920	1300	100	320	117	-	-	Grbeša (2004)
		4700	1000	200	-	117	-	-	Curtin (1983)
		2000-6000	400-2500	100-1500	10-200	27-100	1-18	2-10	Higginbotham & McCarthy (1998)
		4060	590	185	85	30	-	-	Šušić & Guralj (1965)
		4000*	1200*	300*	150*	115	34	18	Olbrich (1963)

*Expressed as oxide salts (K_2O, Na_2O, MgO, CaO).

Table 4. The vitamin content in molasses

Molasses type	Vitamins								Reference
	B7	B9	B5*	B6	B2	B1	B3		
							as niacin	as amide	
	mg/kg								
Cane	1.2-3.2	0.04	54-64	2.6-5	2.5	1.8	300-800	-	Higginbotham & McCarthy (1998)
Cane (blackstrap)	-	-	-	-	trace	trace	14	-	Hickenbottom (1996)
Beet	-	0.1	0.8	2.7	0.8	0.4	29	0.5	Šušić & Sinobad (1989)
Beet	0.04-0.13	0.2	50-100	5	0.4	1.3	400-600	-	Higginbotham & McCarthy (1998)

* In the form of Ca salt.

Apart from macroelements, molasses contains a range of trace elements such as Fe, Mn, Zn, Cu, Co, Se, Bo, Mo, Sr, etc. (Higginbotham & McCarthy, 1998; Olbrich, 1963). The presence of microelements in molasses depends on their occurence in beet root/cane and soil conditions although the exact patterns of their presence and variability have not been fully revealed (Olbrich, 1963). In cane molasses, higher mineral content exists in molasses with higher extractions, i.e., final molasses (Hickenbottom, 1996). Iron is the dominating microelement in both molasses: the average contents are 200 mg/kg in cane and 30 mg/kg in beet molasses (Higginbotham & McCarthy, 1998). According to Higginbotham and McCarthy (1998), iron content in molasses mostly originates from contact with steel equipment in the sugar factory. Mineral content of different molasses types is displayed in Table 3.

In molasses, vitamins stable to heat and in alkaline environments can concentrate. These are mostly B group vitamins. Higginbotham and McCarthy (1998) and Šušić and Sinobad (1989) claimed that vitamins in cane and beet molasses are present in fully bioavailable form which is an additionally advantageous fact that support the benefits of its use in human consumption. Table 4 lists the vitamin content in different types of molasses. Cane molasses contains notable amounts of biotin unlike beet molasses. On the other hand, beet molasses is generally higher in pantothenic acid.

3.3. Health Potential and Antioxidant Capacity

In folk medicine blackstrap molasses has been long administered as a natural remedy for a number of conditions: premenstrual symptom, stress, type 2 diabetes, arthritis, inflammation, iron deficiency, constipation, hair issues, acne and skin conditions and mental symptoms like mood swings. Although little of these health effects have clear scientific substantiation, some effects can be attributed to favorable mineral and vitamin composition and bioavailability. Moreover, recent studies revealed that molasses has impressive antioxidant capacity which probably contributes to the experienced health potential and bioactivity. Recent study reported that filtered molasses concentrate from sugar cane reduced the post-prandial glucose response in human subjects by 5-20% when incorporated as a functional ingredient in carbohydrate-rich food (Wright, et al., 2014). The active compounds for this effect remained unknown but phenolic compounds, minerals (Ca, Mg, K, Cr) and organic acids and/or their synergistic action were supposed to be responsible for the antidiabetic effect.

Table 5. Phenolic profile of raw materials and products from sugar refining industry

Substrate	Phenolic compounds identified	Reference
Sugarcane top	caffeic acid cis-p-hydroxycinnamic acid quercetin apigenin albanin A australone A moracin M 5'-geranyl-5,7,2',4'-tetrahydroxyflavone	Sun, He, Zhao, Li, & Dong (2014)
Sugarcane: *leaves* *juice*	 flavonoid aglycone and glycosides ferulic acid cumaric acid quercetin, caffeic acid ellagic acid	Abbas, Sabir, Ahmad, Boligon, & Athayde (2014)

Substrate	Phenolic compounds identified	Reference
Sugarcane bagasse	tricin 4-O-guaiacylglyceryl ether-7-O-glucopyranoside genistin p-coumaric acid quercetin genistein	Zheng, et al. (2017)
Sugarcane stillage	derivates of caffeoylquinic acids p-hydroxybenzoic acid derivates of feruoyquinic acid vanillic acid syringic acid ferulic acid	Caderby, et al. (2013)
Sugarcane molasses (*antioxidant-rich fraction*)	schaftoside isoschaftoside ferulic acid p-coumaric acid p-hydroxybenzaldehyde p-hydroxyacetophenone	Asikin, et al. (2013)
Cane molasses (*methanolic extract*)	apigenin luteolin tricin chlorogenic acid coumaric acid ferulic acid	Duarte-Almeida, Salatino, Genovese, & Lajolo (2011)
Brown cane sugar	1-methyl-2-pyrrolidinone, 2,3-butanediol, 4-hydroxybenzaldehyde, benzophenone, benzyl alcohol dimethyl sulfoxide syringaldehyde vanillin, acetovanillone benzoic acid dihydro-4-hydroxyfuran-2(3H)-one dodecanoic acid vanillic acid	Payet, Shum-Cheong-Sing, & Smadja (2005)
Cane sugar products: *clear juices, syrups, massecuite, A, B, C molasses*	protocatechuic acid p-hydroxybenzoic acid chlorogenic acid (detected only in molasses) vanillic acid (not detected in clear juice) caffeic acid syringic acid vanillin p-coumaric acid ferulic acid benzoic acid	Payet, Shum-Cheong-Sing, & Smadja (2006)

Table 5. (Continued)

Substrate	Phenolic compounds identified	Reference
A sugar	protocatechuic acid caffeic acid p-coumaric acid ferulic acid	
Commercially available unrefined cane sugar products *AUNO® sugar, sugarcane sugar, black sugar, brown raw sugar*	2,4-di-tert-butylphenol and other	Lee, et al. (2018)
Sugar beet molasses	gallic acid vanillin hydroxybenzoic acid syringic acid cyanidin-3-o-rutinoside delphinidin-3-o-rutinoside delphinidin-3-o-glucuronide ferulic acid	Chen, Zhao, & Yu (2015)
Sugar beet molasses	quercetin 3-o-glucosyl-xyloxide feruoylquinic acid ferulic acid hydroxybenzaldehyde hydroxybenzoic acid luteolin/kaempferol feruloyl-arabinose caffeoyltartaric acid	Valli, et al. (2012)
Sugar beet molasses	gallic acid protocatheuic acid p-hydroxybenzoic acid salicylic acid vanillic acid syringic acid p-coumaric acid ferulic acid protocatheuic aci p-hydroxybenzoic acid vanillin guaiacol	Maestro-Durán, Borja, Jiménez, & León (1996)

It appears that the non-sugar part contains substances that have great potential for health benefits. These substances are maintained from the original crop (cane stem or beet root) and accumulate in molasses during sugar refining. Consequently, they are largely present in low-processed sugar products. Indeed, primitive sugar products, i.e., non-centrifugal sugars (NCS) have been known for numerous positive health effects including anticariogenic, antitoxic, cytoprotective, nephroprotective, immunological effects, etc. NCS is also known as whole cane sugar, black sugar or by numerous traditional names depending on where it is made: jaggery and gur (India), kokuto (Japan), panela (Colombia, Venezuela, Equador), rapadura (Brazil, Sri Lanka), raspadura (Cuba, Panama, Equador), piloncillo (Mexico, Spain), chancaca (Peru, Bolivia, Chile), nam oy (Laos), tapa de dulce (Costa Rica, Nicaragua), uluru dust (Australia), etc. NCS is minimally processed and obtained by concentrating pressed cane juice. Due to minimal purification, reduced to removal of coarse particles and precipitation of proteins and pectin as well as absence of centrifugation, NCS contains nearly all the nonsugars from sugarcane. There is growing scientific evidence to confirm the medicinal effects of NCS and sugarcane extracts. Sugarcane juice has been widely used in Indian traditional medicinal practice to treat various medical conditions such as jaundice, hemorrhage, urinary diseases, skin infections, hearth disorders and its pharmacological action was evidenced in in vitro and animal models (Singh, et al., 2015). Jaffé (2012) reviewed the publications dealing with the health effects of jaggery and found that most frequent reports were associated with nutritional, immunological, anti-toxicity, cytoprotective, anticariogenic effects, antidiabetes and antihypertension effects. The most recent study by Lee et al. (2018) reported on high cellular antioxidant activity in commercial black sugars. The strongest evidence on the medicinal effect of NCS in humans was related to increased hemoglobin and erythrocyte formation due to appreciable iron content (Arcanjo, Pinto, Arcanjo, & Amancio, 2009; Olivares, et al., 2007). Other effects have been observed in vitro or in animal model studies. The therapeutic and medicinal action of NCS or sugarcane extract have been associated to various bioactive compounds and their synergistic actions: cytoprotective and antitoxic effects to antioxidants; anticariogenic effects to the synergistic action of phosphates and two phenolic bioactive compounds isolated; anticarcinogenic effect to glycosides; antiatherosclerotic and immunological effects to phenolics; antidiabetic effects to specific glucosides, minerals and organic acids, etc.

Molasses has strong antioxidative activity and a complex phenolic composition. Previous studies has assessed the antioxidant activity and identified the phenolic profile of sugarcane (Abbas, et al., 2014; Sun, et al.,

2014); various sugar-derived products (Lee, et al., 2018; Singh, et al., 2015; Jaffé, 2015; Okabe, et al., 2009; Payet, Shum-Cheong-Sing, & Smadja, 2005; Nakasone, et al., 1996; Takara, et al., 2003), cane molasses (Payet, Shum-Cheong-Sing, & Smadja, 2006; Duarte-Almeida, Salatino, Genovese, & Lajolo, 2011; Li, et al., 2011), bagasse (Zheng, et al., 2017) and stillage (Caderby, et al., 2013). Table 5 displays the phenolics detected in molasses, by-products and intermediate products of sugar refining as well as in raw materials. Less refined sugar products and products with higher extraction rates had higher antioxidant activity than refined sugar (Lee, et al., 2018; Phillips, Carlsen, & Blomhoff, 2009). There are a limited number of studies focusing on the in vitro antioxidative properties of molasses. Some cane molasses fractions exhibited protective action against oxidative DNA damage (Asikin, et al., 2013). Chen et al. (2015) suggested that sugar beet molasses contains natural antioxidants with significant antitumor activity exhibited in human colon, hepatocellular and breast carcinoma cell lines. Molasses exhibited higher antioxidative activity than honey which had intermediate activity (Phillips, Carlsen, & Blomhoff, 2009). Valli et al. (2012) compared cane and beet molasses in relation to their antioxidant capacity and ability to protect hepatocellular cells from induced oxidative stress and concluded that cane molasses was more effective, exerting even better protective action than α-tocopherol (Valli, et al., 2012). The possible explanation of the weaker antioxidative activity of beet molasses was the different structure of antioxidant constituents. It appears that cane molasses contains more phenolic compounds originating from the plant and associated with polysaccharides whereas beet molasses has less phenolics but more products of alkaline and thermal degradation of sugars and Maillard reactions (Valli, et al., 2012). It seems that the concentration and nature of antioxidative compounds has a major impact on the antioxidative potential of molasses. Compounds responsible for the antioxidant action in molasses are phytocompounds (phenolics, flavonoids, etc.) and products of thermal degradation formed during processing (Maillard reaction products, caramelization products, hidroxymethyl furfural, etc.).

3.4. Other Relevant Components

Other components that may contribute to the functional properties of molasses are organic acids, amino acids and betaine (in beet molasses only). Organic acids present in molasses are displayed in Figure 3. Cane molasses contains 2-4% aconitic acid, up to 1% lactic acid and six to seven organic acids in traces (Higginbotham & McCarthy, 1998). Beet molasses contains around 4% organic acids, mostly lactic acid and was not reported to contain aconitic acid (Higginbotham & McCarthy, 1998). Malic, citric, fumaric, valeric and oxalic acids are present in smaller amounts. Prevalence of volatile acids indicates low quality of molasses (Higginbotham & McCarthy, 1998).

Amino acids are present in free and bound form. Dominant amino acid in cane molasses is aspartic acid whereas glutamic acid prevails in beet molasses (Higginbotham & McCarthy, 1998; Šušić & Sinobad, 1989). Glutamine is degraded during processing to pyrrolidone carboxylic acid (Higginbotham & McCarthy, 1998). Amino acids contribute to color formation in molasses.

Betaine is a valuable nutrient that has been used in fish feeding and pharmaceutical industry. However, recent studies pointed out the importance of betaine for human health. Betaine was found effective in controlling the levels of serum homocysteine, consequently contributing to lower risk from cardiovascular diseases, cancer, neuropathy etc. (Craig, 2004). Dietary intakes of minimum 1500 mg betaine are necessary for the health effect (European Commission (EC), 2012). Food product can bear the health claim related to the positive effect on homocysteine metabolism if its portion is capable of providing at least 500 mg betaine according to the EU Commission Regulation No 432/2012 (European Commission (EC), 2012). Betaine is present only in beet molasses at relatively high level (3-5% w/w, dry substance) (Higginbotham & McCarthy, 1998) that allows its commercial separation in the process of desugarization.

3.5. Contaminants, Processing Chemicals and Toxins

Similarly to any food product, molasses can contain residues of contaminants as the consequence of use of agricultural chemicals and processing aids as well as due to environmental pollution. The reported scientific data showed that molasses rarely contains harmful residues above maximum permitted level. According to data overviewed by Higginbotham & McCarthy (1998) and Filipčev and Lević (2014), both cane and beet molasses generally have low level of heavy elements and PCBs whereas the residues of organochlorine and organophosphorous pesticides rarely exceed 10% of the maximum permitted residue. Jaffé (2015) reported on low level contamination with toxic trace elements and the presence of polycyclic aromatic hydrocarbons in NCS. The type of contaminant depends on the source and growing conditions. It seems that molasses is not at higher risk to contain residues of environmental pollutants as any other food component. On the other hand, molasses may contain traces of process chemicals like sulfur dioxide or surfactants. The permitted level of sulfur dioxide is set in national regulations (for example, max. 70 mg/kg SO_2 for beet molasses in Serbia, max. 200-250 mg/kg total sulfites in USA for feed grade molasses (USDA, 1959)). According to data overview made by Jevtić-Mučibabić (2005) for three consecutive years in the last decade of the 20th century, SO_2 content in beet molasses produced in Serbia spanned over the range 10-50 mg/kg. The presence of surfactants in molasses is usually an issue for fermentation industry due to their detrimental impact on fermentation microorganisms. Molasses normally does not contain residues of surfactants in harmful amounts and problems arise sporadically in the case of excessive use of these substances (Higginbotham & McCarthy, 1998). Little data exist on the mycotoxin contamination of molasses. Abdallah et al. (2016) studied the occurrence of aflatoxin B1 and G1 in sugarcane grass and juice from Egypt. Eighteen percent of juice samples was contaminated with AFG1 whereas 58% of juice contained AFB1 (Abdallah, Krska, & Sulyok, 2016). Cane juice contained less aflatoxin residues than the grass. Hariprasad et al. (2015) reported that sugarcane can uptake aflatoxins from contaminated soil but the contamination was not carried over to jaggery produced from the contaminated samples.

Due to rather severe heat treatments during sugar refining process, molasses contains 5-hydroxymethylfurfural (HMF), formed from reaction of invert sugar with available amino acids. Typical HMF levels in beet and cane molasses are up to 20 and 40 mg/kg, respectively (Higginbotham & McCarthy, 1998). The recent acrylamide survey in selected carbohydrate foods performed by the Canadian Food Inspection Agency revealed that cane molasses had the highest average acrylamide level, 901 μg/kg (Canadian Food Inspection Agency, 2011). However, this level was determined to be unlikely to pose a human health concern.

Besides harmful residues from the environment and processing, molasses can contain potentially detrimental naturally occurring substances like saponins in beet molasses. Beet molasses contains approximately 0.025% saponins which does not pose a health risk, given that potato is 4 times higher in saponine (Filipčev & Lević, 2014).

4. FOOD USES OF MOLASSES

Food industry may benefit from molasses owing mostly to its remarkable mineral composition and outstanding antioxidative capacity. In addition, molasses is an abundant source of other potentially valuable bioactive compounds. Due to high content of dry substance and abundance and good bioavailability of natural bioactive components, supplementation of food products with molasses offer considerable nutritional improvements and health benefits. Food uses of molasses can be manifold; either as a whole ingredient in food manufacture or as a source to isolate functional component(s) or fraction(s) rich in the component of interest which can be further used in formulating functional foods. Chen, Zhao, and Yu (2015) and Guan et al. (2014) agreed upon feasibility of industrial exploitation of natural antioxidants from molasses. Moreover, molasses can be exploited by biotechnological processing techniques as a low-cost raw material for a range of high-value biomolecules such as organic acids, astaxanthin, erythritol, prebiotic oligosacchardies, etc. Sharma et al. (2016) and Lata et al. (2018) proposed cane molasses as a suitable raw material for production of prebiotic and functional bioproducts such as kojibiose, D-allulose, and dextransucrose which are promising ingredients for functional food formulations.

4.1. Use of Molasses in Bakery Products

In bakery industry, cane molasses is used chiefly as a sweetener. Several attributes stands in favor of molasses over other sweeteners: it is natural, nutritious, has a unique flavor and represents an alternative to unpopular highly refined sugars (Hickenbottom, 1996). Bakers use different types of cane molasses at variable levels, depending on the type of product. Hence, cane molasses can be found as an ingredient in several categories of bakery products: breads/rolls, bagels, pies, cakes, cookies, crackers and bakery mixes (Hickenbottom, 1996). The range of application levels is relatively wide, 0.5-25% flour basis (f.b.), depending on the product and molasses type (Hickenbottom, 1996). Hickenbottom (1996) reported on additions up to 6% f.b. in bread, whereas higher levels (≥10%) are usual in formulations of oat and ginger cookies. Molasses is used in bakery applications to enhance sweetness, improve flavor and color of the product. It effectively masks the unappealing greyish nuance of crumb in whole wheat bread, cracked wheat and bran-enriched products (Hickenbottom, 1996). Sweeter and mild-tasting molasses types are used for products with lighter texture of crust and crumb whereas products requiring dark color and strong flavor can benefit from molasses with higher extraction like Third Molasses. Molasses is advantageous in high-fiber products to improve their unattractive flavor. It is frequently used as a natural flavor enhancer compatible with caramel, chocolate, mocha, coffee, cocoa, maple, butterscotch, vanilla and peanut butter (Hickenbottom, 1996). Apart from the role of sweetness, color and flavor enhancer, molasses can be used as a leavening and buffering agent or as a humectant (Hickenbottom, 1996). Owing to its acidity, molasses can improve the leavening effect of soda (Hickenbottom, 1996). Acids in molasses react with soda, releases gas and positively affect the spread of cookies during baking. Molasses also has a buffering capacity due to presence of acids and inorganic salts which is favorable for the control of pH in fermenting systems and stability of acid-sensitive colors (Hickenbottom, 1996). Molasses exerted shelf-life improving effect in intermediate moisture baked products (granola cookies, brownies, fruit cakes, etc.) due to its humectancy and low water activity by absorbing and bonding water (Hickenbottom, 1996). Molasses was found to retard oxidation of high-fat baked systems, being most efficient when added at around 3%, based on total fat (Hickenbottom, 1996).

Table 6. List of baked products supplemented with molasses and the reported effects

Product	Molasses type and dose	Effects	Reference
Wheat bread	Beet molasses (5-10% f.b.)	Improved mineral content (compared to control): • Increase in K by 89% and167%, Ca by 23% and 49%, Mg by 22% and 40%,; • decrease in Na by 0.3% and 0.9%; Increased antiox. potential by 24% and 42%; Decreased specific volume; Increased crumb firmness and decreased resilience, most prominent at 10% level.	Filipčev, et al. (2010); Filipčev (2011)
Gingerbread type biscuits	Beet molasses (10-40% f.b. which equals to 25-100% honey replacement with molasses)	Prominent increase in K, Ca, Mg and Fe content; Relative bioavailability of Fe 26.58-39.37% (decreased by 30%); Rel. bioavailability of Ca 28% (increased by 20%); Increased biscuit spread; Increased flavor intensity; Increased bitter/burnt aftertaste (acceptable up to 20% subst. level); Good overall acceptance.	Filipčev, Bodroža-Solarov, Šimurina, & Cvetković (2012)
Gingerbread type biscuits	Beet molasses (10-20% f.b.)	Significantly increased K, Ca, Mg, Fe content; Significantly decreased biscuit lightness, yellow tone and color vividness, increased red tone; Did not affect changes in hardness and fracturability during 2 month-storage.	Filipčev, Šimurina, & Bodroža-Solarov (2014)

Table 6. (Continued)

Product	Molasses type and dose	Effects	Reference
High-carbohydrate food (white bread, energy bar, wheat flake cereal bricks)	Cane molasses -filtered concentrate of unrefined mill molasses (2.5% in bread, 2% in bar, 1.4-2%in bricks)	Reduction in glycaemic index of products (20% in bread, 11% in bars; 5-8% in cereal bricks) Assumed active compounds responsible for the action: phenolics, minerals, organic acids or synergistic activity between some components.	Wright, Ellis, & Ilag (2014)
Gluten-free biscuit and biscuit dough	Beet molasses liquid and dry (10-50% f.b.)	Weakened dough structure, soft, less elastic and more adhesive dough; Dough with dry molasses easier to handle; Liquid molasses at 40-50% induced higher spread than dry did; Liquid molasses tended to soften crumb, dry molasses hardened it; Increased fracturability especially with dry molasses; Liquid molasses gave better color properties than dry did; Different structure-forming behavior of liquid vs. dry molasses.	Filipčev, et al. (2015)
Gluten-free cookies	Beet molasses liquid and dry (10-40% f.b.)	Improved content of micronutrients (K, Mg, Ca, Fe, betaine); Higher content of antioxidants; Remarkably higher antioxidant activity by DPPH test; Wider spectra of phenolic compounds (dominating were catechin, ferulic and vanillic acid); HMF increased but not to level uncommon to cookies; Dry molasses exerted slightly higher antioxidant activity and total phenolics.	Filipčev, Mišan, Šarić, & Šimurina (2016)

Beet molasses is usually not used for human consumption because of its characteristic, intensive earthy and beety flavor. However, recent research works showed that it is possible to include beet molasses in some food products without compromising their palatability, at levels which provide notable improvements in their nutritional profile and functionality. Successful incorporation of beet molasses was reported in various baked products like bread, ginger-nut biscuits and cookies with or without gluten.

Table 6 summarizes the effects of molasses addition to baked products and ranges of applied doses.

Filipčev noted that white bread supplemented with beet molasses at 5-10% f.b. level had similar, i.e., higher content of macroelements (K, Ca, Mg), respectively, in comparison to Graham bread which supports the value-adding potential of molasses as a mineral enhancer in bakery products (Filipčev, 2011). As already mentioned above, some authors claimed that minerals and vitamins in molasses are in highly bioavailable form (Higginbotham & McCarthy, 1998; Šušić & Sinobad, 1989). Filipčev et al. (2012) examined the relative bioavailability of Ca and Fe in molasses-supplemented gingerbread biscuits and found that the relative bioavailability of Ca was higher by ≈20% in comparison to the control biscuit. However, the Fe relative bioavailability declined by ≈30% in the molasses-supplemented biscuits, presumably due to the presence of potent chelators of Fe like Ca and HMF. Yet, due to relatively high contribution of Fe by molasses, the gingerbread biscuits were similar contributors of available non-heme iron like wholegrain biscuits and amaranth-enriched biscuits (Filipčev, Bodroža-Solarov, Šimurina, & Cvetković, 2012). Beet molasses contains betaine in appreciable amounts. Molasses inclusion to gluten-free cookie formulation increased the betaine content 24-80 times, depending on supplementation level (Filipčev, Mišan, Šarić, & Šimurina, 2016). Molasses-supplemented cookies can serve as a good base for further supplementation with betaine in order to achieve a cookie formulation that would be able to provide 500 mg of betaine per portion and bear a health claim related to the impact of betaine on the metabolism of homocysteine. The study of Wright et al. (2014) revealed an interesting finding: a direct proportion between the dose of filtered molasses concentrate in carbo-hydrate-rich foods and the reduction of the postprandial glycemic and insulin response, measured after the consumption of these foods. The filtered molasses concentrate was prepared from unrefined mill molasses according to the attended procedure of aqueous filtration and fractionation on ion-exclusion chromatography (Wright, Ellis, & Ilag, 2014). The observation is in line with the long-term antidiabetic effect of NCF which is also manufactured from unrefined cane juice (Jaffé, 2012).

4.2. Use of Molasses in Fruit/Vegetable Products

Studies showed that molasses has a potential to be used in osmotic dehydration of fruits and vegetables. One of the earliest records found was the U.S. Patent, authored by Brandner et al. (1947), which mentioned the suitability of cane molasses as a hydrophilic material in a process of "food dehydration by means of hydrophilic liquids" which is essentially alike to the principle of osmotic dehydration. Osmotic dehydration (OD) is a process of moisture reduction in food material (fruits, vegetables, meat) with minimal impact on their quality. It is mostly used as a pretreatment to increase the efficiency of conventional preservation methods such as drying, candying, freezing, freeze/microwave-drying, etc. OD is performed by soaking the plant material in hypertonic (high concentration) solution. Due to differences in osmotic pressure between the food material and solution, water from food tissues migrates to the solution while, simultaneously, solute diffuses to the tissues through cell membranes. In this way, the treated food partially loses water and increases the content of solids. This process is effective under mild ambient conditions and provides better preservation of nutritional, sensory and functional properties of the treated food (Singh, Kumar, & Gupta, 2007). However, since the plant cell membrane responsible for the osmotic mass transport is not perfectly selective, leakage of other solutes occurs from cells into osmotic solution which inevitably leads to some nutritional losses. Substances commonly used for preparation of osmotic solutions are sucrose, sodium chloride and their combinations. Glucose, fructose, maltodextrin and sorbitol are also frequently used as osmotic agents (Yadav & Singh, 2014). The efficiency of OD process can be estimated by the ratio of water loss to solids uptake (WL/SG). Higher values indicate more efficient moisture removal.

Molasses has several interesting features that make it suitable as a hypertonic solution. Firstly, molasses is liquid at high solids content ($\approx$80% w/w) and does not crystallize unlike concentrated sucrose solutions. Secondly, it has high osmotic pressure due to its specific composition which facilitates the diffusion processes. A distinctive advantage of molasses in relation to common hypertonic solutions is its abundance in minerals that results in mineral enrichment of the treated material. From engineering aspect, the main challenge of using molasses in OD applications would be the necessity for high-performance, more expensive pumps to ensure efficient circulation of molasses during dehydration (Šarić, et al., 2016).

Recent studies reported on better efficiency of beet molasses as hypertonic solution in comparison to sucrose solutions during apple treatment (Mišljenović, Koprivica, & Lević, 2010). The treated apples lost more water, gained less sucrose and more minerals when treated in beet molasses (Koprivica, Mišljenović, Lević, & Petkova, 2008). Similar findings were reported for carrots subjected to the similar treatment (Koprivica, Mišljenović, Lević, & Kuljanin, 2009). Molasses can be used as efficient hypertonic solution during several OD cycles; in the case of osmotic dehydration of apples and carrots, molasses could be reused five times after reconditioning that included sieving and sterilization (Mišljenović, Jevrić, Koprivica, & Šobot, 2009). OD of white cabbage in undiluted molasses contributed to 40% increase in K and about 6% increase in Ca and Mg but did not increase the iron content (Cvetković, Jokić, Lević, & Kevrešan, 2013). OD treatment of nettle leaves in beet molasses increased their antioxidant activity in contrast to the treatment in aqueous solution of NaCl and sucrose (Knežević, et al., 2015). Moreover, use of molasses to treat nettle resulted in highest water loss and lowest solid uptake (Knežević, et al., 2015). Optimization studies performed on various fruit and vegetable substrates (apples, plum, carrots, red cabbage, celery leaves and root) showed that, generally, the optimal OD operating conditions included the use of undiluted molasses, immersion times in the range 3-5 h and temperature between 45 and 55°C (Mišljenović, Koprivica, & Lević, 2010; Mišljenović, et al., 2012; Koprivica, et al., 2014; Nićetin, et al., 2015). A potentially adverse impact of molasses as osmotic agent is intensive darkening of the treated samples associated with the transfer of colored compounds from molasses to plant tissue (Šarić, et al., 2016). Moreover, OD results in changes of physical, mechanical and structural features of the plant tissue due to mass and heat transfer gradients (Mayor, Pissara, & Sereno, 2008). Diffusion of water out of plant cell causes plasmolysis which is reflected as shrinkage of vacuoles and protoplasm accompanied with deformation and collapse of cell wall. During OD treatments of apples with molasses and sucrose solutions, Koprivica (2013) observed tissue softening during the first hour of immersion followed by a steady increase in firmness. OD in molasses provoked more pronounced increase in firmness of apple tissue presumably due to incorporation of Ca ions transferred from molasses. After 5 h of immersion in molasses, apple tissue nearly restored its initial firmness.

The main disadvantage of treating fruits/vegetables in molasses is related to changes in the color. The treated fruits/vegetables lose their natural color and take over the dark color of molasses. The darkening is most intensive during the first hour of immersion in molasses and is more intensive on higher temperatures (Koprivica, et al., 2011). The color change greatly limits the further use of dehydrated fruits/vegetables but their successful incorporation in bread and bakery products was shown in the study of Filipčev and coworkers (2010) who elaborated the use of OD apples, plums, carrots and red cabbage in fresh and dried form in bread. Apple pieces dehydrated in beet molasses were successfully applied in fruit yoghurt formulation at 10% supplementation level. OD apples improved the sensory properties of yoghurt by decreasing its acidity and improved the mineral composition (Nićetin, et al., 2014).

4.3. Use of Molasses in Meat Products

A technique of soaking meat in concentrated solution of salt, sugar, acids and seasonings, principally based on osmotic dehydration, has been commonly applied in preservation of meat and meat products. Recent studies reported that beet molasses improves the OD process in meat when used as an osmotic medium, single or in combination with other compounds. Higher efficiency of undiluted molasses as OD agent in dehydration of pork meat, in comparison to common ternary solution (aqueous solution of salt and sucrose), was confirmed in the studies of Filipović, et al. (2012). Moreover, use of molasses provided a sufficient drop in a_w value of treated pork meat, resulting in higher reduction of microorganisms and increasing the hygienic safety of the process (Filipović, et al., 2012). Optimal processing conditions involved pork meat dehydration in undiluted molasses at 22°C (Filipović, et al., 2012). Pork meat dehydrated in beet molasses had higher content of K, Na, Mg, Ca and Fe (Nićetin, et al., 2012). OD treatment of pork meat in molasses was able to preserve microbiological stability of refrigerated pork meat (4°C) for 4 weeks (Šuput, et al., 2013). Combination of OD treatment in molasses, refrigeration and modified atmosphere packaging increased the shelf-life of pork meat up to 6 weeks (Šuput, et al., 2013).

Molasses was suggested to be a suitable alternative for conventional osmotic media in preservation of fish meat which provided better microbiological profile due to efficient lowering of a_w and moisture content (Lončar, et al., 2014). Optimal processing parameters for the best quality of OD crucian carp were immersion for 5 h in undiluted molasses at 35°C (Ćurčić, et al., 2012). Microbiological safety, overall efficiency of the OD process and nutritional improvement was also shown in the treatment of gibel carp meat in molasses (Gubić, et al., 2014).

CONCLUSION

As it can be concluded from the published works included in this chapter, both cane and beet molasses represent concentrates of natural bioactive compounds and micronutrients from cane or beet. Their presence and synergistic action give molasses great potential for use in food as a functional ingredient (in natural form or in fractioned form) to increase the beneficial health effects of food. If not used as a food ingredient, molasses can be used as a raw material for production of other specific functional compounds.

As a single ingredient, molasses offers improvements in many aspects of food quality such as micronutrient pattern (especially due to abundance in potassium and iron, as well as in betaine (beet molasses only)), antioxidant activity and sensory properties. In addition, molasses is affordable and relatively cheap material. The overall effectiveness of food enrichment with molasses is limited due to the necessity for a compromise between sensory properties, nutritional value, and functionality. This is somewhat more emphasized in the case of beet molasses rather than in cane molasses owing to distinctive taste of beet molasses. Due to high content of sugar, sweetening is the primary role of molasses as an ingredient in food formulations which also may limit its application. Nevertheless, molasses is a promising ingredient which practical application in value-added food should not be overlooked. The major food processing sectors that can benefit from the use of molasses are bakery, beverage, fruit/vegetable and meat processing industry.

ACKNOWLEDGMENT

This work was supported by the Ministry of Education, Science and Technological Development of Serbia (grant TR 31055).

REFERENCES

Abbas, S., Sabir, S., Ahmad, S., Boligon, A., and Athayde, M. 2014. "Phenolic profile, antioxidant potential and DNA damage protecting activities of sugarcane *(Saccharum officinarum).*" *Food Chemistry* 147: 10-6.

Abdallah, M., Krska, R., and Sulyok, M. 2016. "Mycotoxin contamination in sugarcane grass and juice: First report on detection multiple mycotoxins and exposure assessment for aflatoxins B1 and G1 in humans." *Toxins* 8: 343-55.

Arcanjo, F., Pinto, V., Arcanjo, M., and Amancio, O. 2009. "Effect of beverage fortified with evaporated sugarcane juice on hemoglobin levels in preschool children." *Revista Panamericana de Salud Pública* 26: 350-54.

Asikin, Y., Takahashi, M., Mishima, T., Mizu, M., Takara, K., and Wada, K. 2013. "Antioxidant activity of sugarcane molasses against 2,2'-

azobis(2-amidino-propane) dihydrochloride-induced peroxyl radicals." *Food Chemistry* 141: 466-72.

Brandner, J., Wilmington, and Goepp Jr, R. 1947. Patent No. 2,420,517. USA.

Bruhns, G. 1998. "Historical overview of sweeteners." In Sugar Technology. *Beet and Cane Sugar Manufacture*, edited by P. van der Poel, H. Schiweck, and T. Schwartz, 1-53. Berlin: Verlag Dr. Albert Bartens KG.

Caderby, E., Baumberger, S., Hoareau, W., Fargues, C., Decloux, M., and Maillard, M.-N. 2013. "Sugar cane stillage: A potential source of natural antioxidants." *Journal of Agricultural and Food Chemistry* 61: 11494-501.

Canadian Food Inspection Agency. 2011. "*Acrylamide in Selected Foods-Archived 2010-2011.*" Accessed August 1, 2018. http://www.inspection.gc.ca/food/chemical-residues-microbiology/chemical-residues/2010-2011-acrylamide/eng/1394128352316/1394128433628

Chen, M., Zhao, Y., and Yu, S. 2015. "Optimisation of ultrasonic-assisted extraction of phenolic compounds, antioxidants, and antho-cyanins from sugar beet molasses." *Food Chemistry* 172: 543-50.

Clarke, M. 1993. "Syrups". In *Encyclopaedia of Food Science, Food Technology and Nutrition*, edited by R. Macrae, Richard Kenneth Robinson, Michèle J. Sadler, 5711-17. Cambridge, Massachusetts: Academic Press.

Craig, S. A. 2004. "Betaine in human nutrition." *American Journal of Clinical Nutrition* 80: 539-48.

Ćurčić, B., Lević, L., Filipović, V., Nićetin, M., Knežević, V., Pezo, L., and Šuput, D. 2012. "Osmotic drying of crucian carp *(Carassius carassius)* using sugar beet molasses solutions." *Journal of Processing and Engineering in Agriculture* 16(4): 173-5.

Curtin, L. 1983. *Molasses-General Consideration.* West des Moines, Iowa: National Feed Ingredients Association.

Cvetković, B., Jokić, A., Lević, L., and Kevrešan, Ž. 2013. "Osmotic dehydration of white cabbage in different hypertonic solutions-mass transfer kinetics and improvement of nutritional value of the developed product." *Journal on Processing and Energy in Agriculture* 17: 154-7.

Duarte-Almeida, J., Salatino, A., Genovese, M., and Lajolo, F. 2011. "Phenolic composition and antioxidant activity of culms and sugarcane (*Saccharum officinarum* L.) products." *Food Chemistry* 125: 660-4.

European Commission (EC). 2012. "Commission Regulation (EU) No 432/2012 of 16 May 2012 establishing a list of permitted health claims made on foods, other than those referring to the reduction of disease risk and to children's development and health." *Official Journal of the European Union* L: 136(55), 1-40.

Filipčev, B. 2011. "Sugar beet molasses: An ingredient to enhance micronutrients and functionality in bread." *Food and Feed Research* 38(2): 95-100.

Filipčev, B., and Lević, L. 2014. *Primena melase šećerne repe i poluproizvoda od osmotski dehidriranog voća/povrća u melasi kao nutritivno vrednih dodataka za obogaćivanje pekarskih i finih pekarskih proizvoda.* Novi Sad, Vojvodina, Srbija: Univerzitet u Novom Sadu, Institut za prehrambene tehnologije. [*Use of Sugar Beet Molasses and Semi-Products of Osmotic Dehydration of Fruits/Vegetables in Molasses as Nutritionally Valuable Ingredients to Enhance Bakery and Fine Bakery Products*. Novi Sad, Vojvodina, Serbia: University of Novi Sad, Institute of Food Technology].

Filipčev, B., Bodroža-Solarov, M., Šimurina, O., and Cvetković, B. 2012. "Use of sugar beet molasses in processing of gingerbread type biscuits: effect on quality characteristics, nutritional profile, and bioavailability of calcium and iron." *Acta Alimentaria* 41(4): 494-505.

Filipčev, B., Lević, L., Bodroža-Solarov, M., Mišljenović, N., and Koprivica, G. 2010. "Quality characteristics and antioxidant properties of breads supplemented with sugar beet molasses-based ingredients." *International Journal of Food Properties* 13: 1035-53.

Filipčev, B., Mišan, A., Šarić, B., and Šimurina, O. 2016. "Sugar beet molasses as an ingredient to enhance the nutritional and functional properties of gluten-free cookies." *International Journal of Food Sciences and Nutrition* 67(3): 249-56.

Filipčev, B., Šimurina, O., and Bodroža-Solarov, M. 2014. "Quality of gingernut type biscuits as affected by varying fat content and partial replacement of honey with molasses." *Journal of Food Science and Technology* 51(11): 3163-71.

Filipčev, B., Šimurina, O., Dapčević Hadnađev, T., Jevtić-Mučibabić, R., Filipović, V., and Lončar, B. (2015). "Effect of liquid (native) and dry molasses originating from sugar beet on physical and textural properties of gluten-free biscuit and biscuit dough." *Journal of Texture Studies* 46: 353-64.

Filipović, V., Ćurčić, B., Nićetin, M., Plavšić, D., and Koprivica, G. M. 2012. "Mass transfer and microbiological profile of pork meat dehydrated in two different osmotic solutions." *Hemijska Industrija* 66(5): 743-48.

Galanakis, C. 2012. "Recovery of high added-value components from food wastes: Conventional, emerging technologies and commercialized applications." *Trends in Food Science & Technology* 26, 68-87.

Gasmalla, M., Yang, R., Nikoo, M., and Man, S. 2012. "Production of ethanol from Sudanese sugar cane molasses and evaluation of its quality." *Food Processing & Technology* 3(7): 1-3.

Google. 2018a. "*Molasses-Glossary Term*." Accessed June 21, 2018. http://recipetips.com/glossary-term/t--33503/molasses.asp.

Google. 2018b. "*Nutrition Data-Blackstrap Unsulphured Molasses*." Accessed June 21, 2018. http://nutritiondata.self.com/facts/custom/3139363/2.

Grbeša, D. 2004. *Metode procjene i tablice kemijskog sastava i hranljive vrijednosti krepkih krmiva*. Zagreb, Hrvatska: Agronomski fakultet Sveučilišta u Zagrebu, Hrvatsko agronomsko društvo. [*Methods for Estimation and Tables of Chemical Composition and Nutritive Value of Feed*. Zagreb, Croatia: Faculty of Agriculture, University of Zagreb, Croatia].

Guan, Y., Tang, Q., Fu, X., Yu, S., Wu, S., and Chen, M. 2014. "Preparation of antioxidants from sugarcane molasses." *Food Chemistry* 152: 552-7.

Gubić, J., Plavšić, D., Šarić, L., Varga, A., Čabarkapa, I., Filipčev, B., and Šimurina, O. 2014. "Comparative investigation of fish *(Carassius gibelio)* treated by osmotic dehydration in molasses at different temperatures." *Food and Feed Research* 41(2): 109-14.

Haney, R. 2018. "*Types of Molasses.*" Accessed June 13, 2018. http://healthyeating.sfgate.com/types-molasses-2894.html.

Hariprasad, P., Vipin, A., Karuna, S., Raksha, R., and Venkateswaran, G. 2015. "Natural aflatoxin uptake by sugarcane (*Saccharum officinarum* L.) and its persistence in jaggery." *Environmental Science and Pollution Research* 22(8): 6246-53.

Hickenbottom, J. 1996. "Use of molasses in bakery products." *AIB Technical Bulletin* XVIII(6): 1-6.

Higginbotham, J. D., and McCarthy, J. 1998. "Quality and storage of molasses." In *Sugar Technology-Beet and Cane Manufacture*, edited by P. W. van der Poel, H. Schiweck, and T. Schwartz, 973-92. Berlin: Verlag Dr. Albert Bartens KG.

Jaffé, W. 2012. "Health effects of non-centrifugal sugar (NSC): A review." *Sugar Tech* 14(2): 87-94.

Jaffé, W. 2015. "Nutritional and functional components of non centrifugal cane sugar: A compilation of the data from the analytical literature." *Journal of Food Composition and Analysis* 43: 194-202.

Jevtić-Mučibabić, R. 2005. "*Standardizacija melase za fermentacionu industriju.*" Magistarski rad, Univerzitet u Novom Sadu. ["*Molasses Standardization Aiming to Fit the Requirements of Fermentation Industry.*" MSc Thesis, University of Novi Sad].

Jevtić-Mučibabić, R., Grbić, J., Filipčev, B., Šimurina, O., and Kuljanin, T. 2013. "Mineral substances in sugar beet molasses." In *Proceedings of the III International Congress "Engineering, Environment and Materials in processing Industry",* edited by M. Gligorić, A. Došić, D. Keševlj, and D. Vujadinović, 275-80. Zvornik, Bosnia and Herzegovina: University of East Sarajevo, Faculty of Technology.

Knežević, V., Pezo, L., Lončar, B., Nićetin, M., Filipović, V., Gorjanović, S., Sužnjević, D, and Kuljanin, T. 2015. "Osmotic treatment of nettle leaves-optimization of kinetics and antioxidant activity." *Journal of Processing and Energy in Agriculture* 19(4): 175-8.

Koprivica, G. 2013. "*Nutritive profile and sensorial quality of osmotically dehydrated fruits and vegetables in sugar beet molasses and sucrose solutions.*" PhD diss., Faculty of Technology, University of Novi Sad.

Koprivica, G., Lato, P., Ćurčić, B., Lević, L., and Šuput, D. 2014. "Optimization of osmotic dehydration of apples in sugar beet molasses." *Journal of Food Processing and Preservation* 38: 1705-15.

Koprivica, G., Mišljenović, N., Lević, L., Jokanović, M., and Jevrić, L. 2011. "Effect of process parameters on the color of apple osmotically dehydrated in sugar beet molasses." *Journal on Processing and Energy in Agriculture* 15 (4): 223-6.

Koprivica, G., Mišljenović, N., Lević, L., and Kuljanin, T. 2009. "Influence of the nutrients present in sugar beet molasses and saccharose solutions on the quality of osmodehydrated carrots." *Journal on Processing and Energy in Agriculture* 13(2): 188-91.

Koprivica, G., Mišljenović, N., Lević, L., and Petkova, M. 2008. "Osmotic dehydration of apple in sucrose and sugar beet molasses: change of nutritive properties of final product." *Journal on Processing and Energy in Agriculture* 12(4): 215-18.

Lata, K., Sharma, M., Patel, S., Sangwan, R., and Singh, S. 2018. "An integrated bio-process for production of functional biomolecules utilizing raw and by-products from dairy and sugarcane industries." *Bioprocess and Biosystems Engineering* 41(8): 1121-31.

Lee, J. S., Ramalingam, S., Jo, I.G., Kwon, Y. S., Bahuguna, A., Oh, Y. S., Kim, M. 2018. "Comparative study of the physicochemical, nutritional, and antioxidant properties of some commercial refined and non-centrifugal sugars." *Food Research International* 109: 614-25.

Li, X., Song, H., Yao, S., Jia, C., Yang, Y., and Zhu, W. 2011. "Quantitative analysis and recovery optimization of flavonoids and anthocyanins in sugar-making process of sugarcane industry." *Food Chemistry* 125: 150-7.

Lončar, B., Filipović, V., Nićetin, M., Knežević, V., Pezo, L., Plavšić, D., and Šarić, L. 2014. "Microbiological profile of fish dehydrated in two different osmotic solutions." *Acta Universitatis Sapientiae* 7: 73-80.

Madsen, R. 1998. "Juice purification systems." In *Sugar Technology. Beet and Cane Sugar Manufacture* edited by P. van der Poel, H. Schiweck, and T. Schwartz, 500-22. Berlin: Verlag Dr. Albert Bartens KG.

Maestro-Durán, R., Borja, R., Jiménez, A., & León, M. 1996. "Phenolic compounds in sugar beet molasses." *Agrochimica* 40, 173-9.

Mayor, L., Pissara, J., and Sereno, A. M. 2008. "Microstructural changes during osmotic dehydration of pumpkin." *Journal of Food Engineering* 85: 326-39.

Mišljenović, N., Jevrić, l., Koprivica, G. L., Šobot, K. 2009. "*Characteristics of sugar beet molasses reused in process of osmotic dehydration of apple and carrot.*" Paper presented at the annual meeting for the Faculty of Technology in Leskovac, University of Niš, Leskovac, Serbia.

Mišljenović, N., Koprivica, G., and Lević, L. 2010. "Comparison of the kinetics of osmotic drying apples in sugar beet molasses and sucrose solutions." *Journal on Processing and Energy in Agriculture* 14(1): 32-5.

Mišljenović, N., Koprivica, G., Pezo, L., Lević, L., Ćurčić, B., Filipović, V., and Nićetin, M. 2012. "Optimization of the osmotic dehydration of carrot cubes in sugar beet molasses." *Thermal Science* 16(1): 43-52.

Nakasone, Y., Takara, K., Wada, K., Tanaka, J., Yogi, S., and Nakatani, N. 1996. "Antioxidative compounds isolated from Kokuto, non-centrifugal cane sugar." *Bioscience, Biotechnology and Biochemistry* 60: 1714-16.

Nićetin, M., Filipović, V., Ćurčić, B., Pezo, L., Knežević, V., Gubić, J., and Kuljanin, T. 2012. "The influence of different osmotic solutions on nutritive profile during osmotic dehydration of pork." Paper presented at *XV International Feed Technology Symposium,* COST-"Feed for Health" joint Workshop, Novi Sad, Serbia, October 3-5.

Nićetin, M., Lević, L., Lončar, B., Filipović, V., Knežević, V., Kuljanin, T., and Pezo, L. 2014. Characterization of fruit yoghurt with apple cubes osmotically dehydrated in molasses. *Journal of Hygienic Engineering and Design* 9: 64-7.

Nićetin, M., Lončar, B., Filipović, V., Knežević, V., Kuljanin, T., Pezo, L., and Gorjanović, S. 2015. Changes in antioxidative activity and phenolic content of celery leaves and root during osmotic treatment. *Journal of Hygienic Engineering and Design* 13: 75-81.

Okabe, T., Toda, T., Inafuku, M., Wada, K., Iwasaki, H., and Oku, H. 2009. Antiatherosclerotic function of Kokuto, Okinawan noncentri-fugal cane sugar. *Journal of Agricultural and Food Chemistry* 57: 69-75.

Olbrich, H. 1963. *The Molasses*. Berlin: Biotechnologie-Kemp GmbH.

Olivares, M., Pizzaro, F., Hertrampf, E., Fuenmayor, G., and Estevez, E. 2007. "Iron absorption from wheat flour: Effects of lemonade and chamomile infusion." *Nutrition* 23: 296-300.

Payet, B., Shum-Cheong-Sing, A., and Smadja, J. 2005. "Assessment of antioxidants activity of cane brown sugars by ATPS and DPPH radical scavenging assays: Determination of their polyphenolic and volatile constituents." *Journal of Agricultural and Food Chemistry* 53: 10074-9.

Payet, B., Shum-Cheong-Sing, A., and Smadja, J. 2006. "Comparison of the concentrations of phenolic constituents in cane sugar manufacturing products with their antioxidant activities." *Journal of Agricultural and Food Chemistry* 54: 7270-6.

Perez, R. 1995. "Molasses." Paper presented at the First FAO Electronic Conference. *FAO*. Accessed June 15, 2018. www.fao.org/livestock/agap/frg/ECONF95/PDF/MOLASSES.pdf.

Phillips, K., Carlsen, M., and Blomhoff, R. 2009. "Total antioxidant content of alternatives to refined sugar." *Journal of American Dietetic Association* 109(1): 64-71.

Šarić, L., Filipčev, B., Šimurina, O., Plavšić, D., Šarić, B., Lazarević, J., and Milovanović, I. 2016. "Sugar beet molasses: properties and applications in osmotic dehydration of fruits and vegetables." *Food and Feed Research* 43(2): 135-44.

Sauvant, C., Perez, J.-M., and Tran, G. 2004. *Tables of Composition and Nutritional Value of Feed Materials*. Wageningen: Wageningen Academic Publishers.

Serbian Regulation. 2013. "Pravilnik o prehrambenim aditivima." *Službeni glasnik RS*: 63. ["Rulebook on Food Additives." *Official Gazette of RS* 63: 1-164].

Serbian Standard. 1963. SRPS E.L3.020/1963.

Sharma, M., Patel, S., Lata, K., Singh, U., Krishania, M., Sangwan, R., and Singh, S. (2016). "A novel approach of integrated bioprocessing of cane molasses for production of prebiotic and functional bioproducts." *Bioresource Technology* 219: 311-8.

Singh, A., Lal, UR, Mukhtar, H., Singh, P., Shah, G., and Dhawan, R. 2015. "Phytochemical profile of sugarcane and its potential health aspects." *Pharmacognosy Review* 9(17): 45-54.

Singh, B., Kumar, A., and Gupta, A. 2007. "Study of mass transfer kinetics and effective diffusivity during osmotic dehydration of carrot cubes." *Journal of Food Engineering* 79: 471-80.

Sun, J., He, X., Zhao, M., Li, C., and Dong, Y. 2014. "Antioxidant and nitritite-scavenging capacities of phenolic compounds from sugarcane *(Saccharum officinarum* L.*)* tops." *Molecules* 19(9): 13147-60.

Šuput, D., Lazić, V., Pezo, L., Lević, L., Gubić, J., Hromiš, N., and Šojić, B. 2013. "Modified atmosphere packaging and osmotic dehydration effect in pork quality and stability." *Romanian Biotechnological Letters* 18(2): 8160-8.

Šušić, S., and Guralj, E. 1965. *Osnovi tehnologije šećera.* Beograd: Naučna knjiga. [*The Basics of Sugar Technology*. Belgrade: Naučna knjiga].

Šušić, S., and Sinobad, V. 1989. "Ispitivanja u cilju unapređenja industrije šećera Jugoslavije." *Hemijska Industrija* 43: 10-21. ["A research study to advance sugar industry in Yugoslavia." *Chemical Industry* 43: 10-21].

Takara, K., Matsui, D., Wada, K., Ishiba, T., Chinen, I., and Nakasone, Y. 2003. "New phenolic compounds from Kokuto, non-centrifuged cane sugar." *Bioscience, Biotechnology and Biochemistry* 67: 376-79.

USDA. (1959). *United States Standards for Grades of Sugarcane Molasses.* Washington D.C.: Processed Products Branch- Fruit and Vegetable Division, AMS, USDA.

Valli, V., Gómez-Caravaca, A., Di Nunzio, M., Danesi, F., Caboni, M., and Bordoni, A. 2012. "Sugar cane and sugar beet molasses, antioxidant-rich alternatives to refined sugar." *Journal of Agricultural and Food Chemistry* 60: 12508-15.

Wright, A. G., Ellis, T. P., and Ilag, L. L. 2014. "Filtered molasses concentrate from sugar cane: natural functional ingredient effective in lowering the glycaemic index and insulin response of high carbohydrate foods." *Plant Foods for Human Nutrition* 69(4): 310-16.

Yadav, A. K., and Singh, S. V. 2014. "Osmotic dehydration of fruits and vegetables: a review." *Journal of Food Science and Technology* 51(9): 1654-73.

Zheng, R., Su, S., Zhou, H., Yan, H., Ye, J., Zhao, Z., You, L., and Fu, X. 2017. "Antioxidant/antihyperglycemic activity of phenolics from sugarcane *(Saccharum officinarum* L.*)* bagasse and identification by UHPLC-HR-TOFMS." *Industrial Crops and Products* 101: 104-14.

In: Molasses: Forms, Production and Uses ISBN: 978-1-53614-703-2
Editors: K. Maddison and R. Fuller

Chapter 6

THERMOPHILIC BIOMETHANE PRODUCTION BY CO-DIGESTING GLYCERIN AND MOLASSES IN AN AnSBBR: EFFECTS OF COMPOSITION, FEED STRATEGY AND APPLIED ORGANIC LOAD

Natalia F. Zucoloto, Giovanna Lovato, Roberta Albanez, Suzana M. Ratusznei and José A. D. Rodrigues*

Mauá School of Engineering,
Mauá Institute of Technology (EEM/IMT),
São Caetano do Sul, SP, Brazil

* Corresponding Author Email: rodrigues@maua.br.

ABSTRACT

An assessment was made regarding methane production from glycerin digestion and glycerin/molasses co-digestion, under thermophilic conditions (55°C), in a mechanically stirred (50 rpm) anaerobic reactor, operated in sequencing batch and fed-batch, containing immobilized biomass on polyurethane foam (AnSBBR). The performance of the 5.6 L AnSBBR was assessed at increasing applied volumetric organic loading rate (OLRA) and according to the feeding strategy. During mono-digestion the AnSBBR was batch operated with 2.3 to 6.5 gCOD.L^{-1}.d^{-1} and cycle length of 8 h. During co-digestion (86% glycerin and 14% molasses) the AnSBBR was operated in batch mode with OLRA of 2.4 to 7.7 gCOD.L^{-1}.d^{-1} and cycle length of 8 h, and in fed-batch mode with OLRA of 7.7 gCOD.L^{-1}.d^{-1}, cycle length of 8 h and feeding time of 4 h. The best results were obtained during co-digestion with 7.7 gCOD.L^{-1}.d^{-1}. In the batch operation, a molar productivity of 84.4 molCH$_4$.m^{-3}.d^{-1}, methane yield of 11.0 molCH$_4$.g COD^{-1} with 71.6% of methane in the biogas was obtained. In the fed-batch operation, molar productivity of 83.7 molCH$_4$.m^{-3}.d^{-1}, methane yield 11.2 molCH$_4$.gCOD^{-1} was obtained with 69.1% of methane in the biogas. There was no significant difference between the results obtained when the feeding strategy was altered, so co-digestion could be performed in both batch and fed-batch with OLRA of 7.7 gCOD.L^{-1}.d^{-1}. During mono-digestion hydrogenotrophic methanogenesis predominated up to OLRA of 4.3 mgCOD.L^{-1}.d^{-1}. However, when OLRA increased acetoclastic methanogenesis became dominant. During the batch co-digestion hydrogenotrophic methanogenesis was predominant, whereas during the fed-batch co-digestion acetoclastic methanogenesis predominated. The scaling up estimation, considering small and medium-sized biodiesel production industries, resulted in single reactors of 6.64 and 332 m^3, for which parallel operation is suggested with 4 and 8 reactors, with estimated energy generation of 2.94 and 147 kW, respectively.

Keywords: AnSBBR, biomethane, organic load, co-digestion, mono-digestion, glycerin, molasses, thermophilic

1. INTRODUCTION

With the growing concerns related to global warming, the generation of crude glycerol is expected to expand with the increase in biodiesel production (Kurahashi et al., 2017). This increase created a glycerol surplus that resulted in a dramatic decrease in crude glycerol prices. European Union directives, however, determined that the use of biodiesel should reach 20% of the fuels used in 2020 (Maragkaki et al., 2017). In the biodiesel production process, glycerol is the main byproduct, representing approximately 10 wt.% (Nuchdang and Phalakornkule, 2012). Despite its widespread applications in the pharmaceutical industry, in the food and cosmetics industries, glycerol refining to a high degree of purity is very expensive, especially for small and medium-sized biodiesel producers (Maragkaki et al., 2017), in addition to generating large quantities of highly polluted wastewaters (Siles et al., 2010).

Anaerobic digestion is an alternative for revaluing this byproduct of biodiesel production, which can be defined as the biological conversion of organic matter into a variety of final products, including biogas, which is main constituted by methane (65-70%) and carbon dioxide. It has the advantages of low biological sludge production, reduced nutrient requirements, high efficiency and methane production. Thus, it allows pollution control and the recovery of energy in wastes (Siles et al., 2010).

In order to improve the productivity of crude glycerol anaerobic digestion, some studies of glycerol co-digestion with other residues (such as urban solid waste, agroindustry byproducts and slurry) were carried out to improve methane production (Vlassis et al., 2013). In this way, sugar cane molasses stands out as a promising candidate due to its high organic matter content and the possibility of energy recovery by biogas production.

Anaerobic reactors operated in sequential batch with immobilized biomass (AnSBBR) allow the typical biological anaerobic metabolism, from the consumption of substrate to the production of methane and carbon dioxide. They operate according to the following cyclical steps: feeding, reaction and discharge. The feeding step may have a variable filling time, the batch operation is performed when the filling time is negligible compared to the total cycle length, and the fed batch operation is when the filling time is significant compared to cycle length. In the reaction step, the treatment itself occurs, by means of biological transformations of the waste organic matter. In the discharge step, there is the withdrawn of the treated and clarified effluent. The major advantages of this type of operation are its simplicity, the efficient effluent quality control and the possibility of application to a large variety of wastewaters (Lovato et al., 2015). There has been an in increase in the literature about the potential application of these reactors for the bioenergy production. This application occurs in a wide range of wastewaters that have become a raw material instead of a process residue. This maintains the objective of producing, in a viable way, energy from hydrogen and methane, while ensuring the appropriate environmental disposal of all waste (Lovato et al., 2015).

In this context, the main objective of this work was to evaluate the application of an anaerobic reactor operated in sequencing batch and fed batch containing immobilized biomass and mechanical stirring (AnSBBR) for biomethane production by digesting glycerin and co-digesting glycerin and molasses under thermophilic conditions (55°C). The influence of increasing applied volumetric organic load, feeding strategy and the addition of the co-substrate molasses were evaluated on the stability and process performance regarding organic matter removal and biogas productivity and composition. In addition, biokinetic modeling was used to better understand the metabolic pathways of the anaerobic digestion. Finally, a scale-up assessment of the investigated reactor was carried out, to estimate the potentially recovered energy from the produced methane on an industrial scale in order to analyze economic aspects of the technological option proposed in this study.

2. Materials and Methods

2.1. AnSBBR

The AnSBBR (Figure 1) was operated at a temperature of 55ºC, controlled by water circulating in the external wall of the reactor. The temperature was regulated by an ultra-thermostatic bath. Mechanical stirring was set at 100 rpm and the complete cycle was set at 8 h.

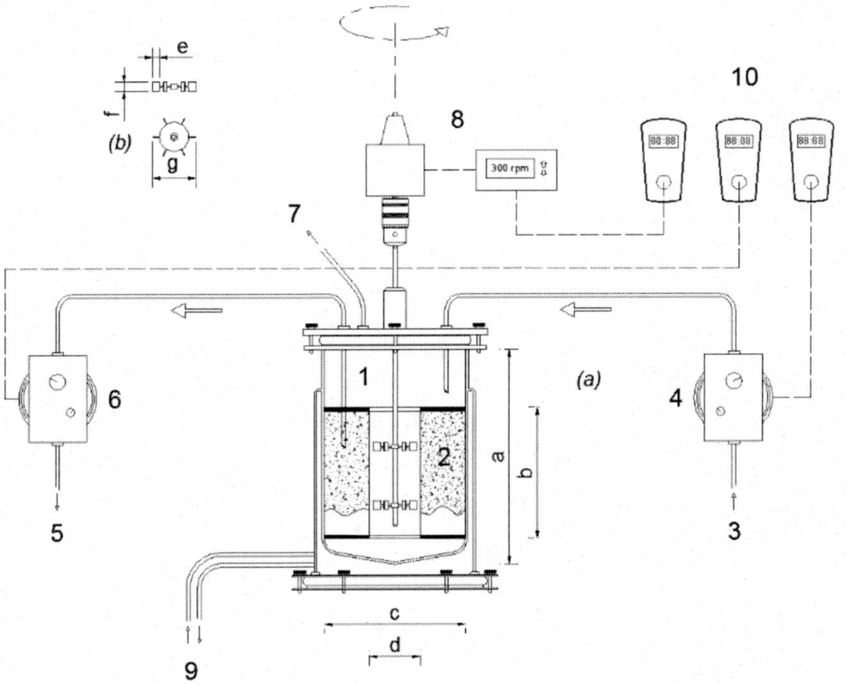

Figure 1. Scheme of the reactor used in this study. [Notation: (a) Reactor BIOFLO III (New Brunswick Scientific Co.) of 6 L capacity; 2 – basket containing support material with immobilized biomass (b = 18.0 cm; d = 7.0 cm); 3 – influent; 4 – feed pump; 5 – Effluent; 6 – Discharge pump; 7 – Biogas outlet; 8 – Mechanical agitation; 9 – Temperature control system (ultra-thermostatic bath); (b) details of the six-flat-blade turbine impellers] (Albanez et al., 2016).

2.2. Wastewater

In the mono-digestion phase, the wastewater used was prepared from double distilled glycerin, which came from an industrial plant of biodiesel production located in São Paulo, Brazil. Its organic matter concentration and density were of 1436 $gCOD.L^{-1}$ and 1292 $g.mL^{-1}$, respectively.

In the co-digestion phase, the wastewater was prepared with glycerin and molasses. Molasses was from the industrial process of bioethanol production of a plant located in São Paulo, Brazil. Its organic matter concentration was 1104 gCOD.L^{-1} and density of 1364 g.mL^{-1}.

In every assay, glycerin was diluted with water so as to achieve the desired organic loading rate. It was also supplemented with urea and sodium bicarbonate.

2.3. Inoculum and Inert Support

The inoculum used came from a thermophilic UASB reactor treating vinasse from a wastewater treatment system of an ethanol plant and presented total volatile solids of 17 g.L^{-1}. The inert support (polyurethane foam cubes) used in all experiments was the same as that from Almeida et al., (2017) and the immobilization process was according to Zaiat et al., (1994).

2.4. Physical-Chemical Analyses and Microbiological Tests

Influent and effluent samples were taken to monitor the reactor behavior. The performance of the reactor was monitored in terms of organic matter (Chemical Oxygen Demand (COD – C_S), Carbohydrate (C_C) in unfiltered (C_{ST} and C_{CT}) and filtered forms (C_{SF} and C_{CF}) and filtered glycerin (C_{GT})). For the filtered samples, micro fiberglass membranes with nominal opening diameter of 0.45 μm were used. The stability of the system was monitored by daily analysis of pH, bicarbonate alkalinity (BA), total volatile acids (TVA), total solids (TS), total volatiles solids (TVS), total suspended solids (TSS) and volatile suspended solids (VSS). Analyses were performed according to Standard Methods for the Examination of Water and Wastewater (APHA/AWWA/WEF 2012), Ripley et al., (1986) for alkalinity determination, Dubois et al. (1956) for carbohydrate determination and an adapted method from Bondioli and Della Bella (2005) for glycerin determination. The intermediate compounds of the anaerobic metabolism (organic acids: ethanol, acetic, propionic, butyric/isobutyric and valeric/

isovaleric) were studied once per assay. The final biogas composition (in terms of carbon dioxide - CO_2 and methane - CH_4) were analyzed daily, as described in Almeida et al. (2017).

For the microbiological test, an inert support sample with biomass was collected and examined on glass slides covered with 2% agar film for standard optical microscopy and fluorescence phase contrast, using a microscopy Olympus® model BX41, with digital camera system *Optronics* and image acquisition by the *Image Pro-Plus*® software version 4.5.0.

2.5. Stability and Performance Indicators

All stability and performance indicators used are described in Almeida et al. (2017) and Albuquerque et al. (2019): removal efficiency of total organic matter for total and soluble samples (ε_T and ε_S), glycerin removal for filtered samples (ε_G) applied volumetric organic load (OLR_A), removed volumetric organic load (OLR_R), volumetric and specific molar productivity (MPr and SMPr) and the yield factor of methane produced per applied load (Y_A) and per removed load (Y_R).

The biogas volume during a cycle (V_G) was measured using a Ritter® MilligasCounter. For the fed-batch condition, the accumulated volumes of biogas at each point were calculated by Equation (1) ($V_{Gi\text{-}BA}$). In this method, the biogas production needs an adjustment because of the influent volume added to the reactor. Equation (2) presents the method for the batch mode condition ($V_{Gi\text{-}B}$), using the data recorded by the biogas measurement ($V_{M\text{-}i}$). The other terms are the biogas profile point number (N_i), the total number of profile points during the fed batch ($N_{t\text{-}BA}$) and the volume of affluent fed during the cycle (V_A).

$$V_{Gi-BA} = V_{M-i} - \frac{N_i}{N_{t-BA}} V_A \tag{1}$$

$$V_{Gi-B} = V_{M-i} - V_A \tag{2}$$

The biogas data obtained were converted to STP (standard temperature/ pressure). The conversion of the biogas volume was calculated according to the general gas law by Equation (3) and the generated methane (n_{CH4} in $mmol.d^{-1}$) was calculated by Equation (4). The volume used at STP and the biogas volume to be converted (V_i - V_{Gi-BA} or V_{Gi-B}). For this method the following were considered: air pressure at the measurement site (P_A), partial pressure of water vapor (P_V), pressure of the liquid column above the measuring chamber (P_L - 2 mbar), standard pressure (P_N - 1013.25 mbar), standard temperature (T_N - 273.15 K), temperature inside the reactor (T_a - 55 °C), pressure (P - 1 atm), volume at "standard" conditions (1 atm/273 K - V_N) and Clapeyron constant (R - 0.082 $atm.L.K^{-1}.mol^{-1}$).

$$V_N = V_i \frac{(P_a - P_V + P_L)}{P_N} \frac{T_N}{T_a} \tag{3}$$

$$n_{CH_4} = \frac{P.V_N}{R.T_N} \tag{4}$$

At the end of each condition, after profile analysis, the volume inside de reactor was drained and measured (V_R). All inert support with the immobilized biomass was weighed (M_{T-SI+B}) and then a sample of this material was collected and quantified (M_{A-SI+B}). From that sample the solid (inert support) and liquid phases (biomass which was immobilized) were separated. With the solid phase, total solids were analyzed. From the liquid phase, TS (M_{A-TS}) and TVS assays (M_{A-TVS}) were carried out. The total amount of biomass of the reactor (M_{TVS}) was calculated by Equation (5). The relation between the quantity of biomass and reactor liquid medium (C_X) was calculated by Equation (6) and the relation between the quantity of biomass and inert support ($C_{X'}$) was calculated by Equation (7).

$$M_{TVS} = \frac{M_{A-TVS} . M_{T-SI+B}}{M_{A-SI+B}} \tag{5}$$

$$C_X = \frac{M_{TVS}}{V_R} \tag{6}$$

$$C_{X'} = \frac{M_{A-TVS}}{M_{A-SI}} \tag{7}$$

2.6. Kinetic Metabolic Model

The kinetic model of macro metabolic pathways proposed in this work was adapted from Lovato et al. (2017). It is important to highlight that the proposed model aims to understand the macro metabolic pathways of the system, and not to predict experimental data and scale-up assessment.

In the adopted kinetic model developed for the anaerobic sequencing batch/fed-batch biofilm reactor, four volatile acids (acetic, propionic, butyric and valeric acids) were considered.

The anaerobic process was simplified in nine steps (Equations 1 to 9). In the first four parallel steps (Equations 1 to 4 - *hydrolysis and acidogenesis*), the substrate (S, which is considered the molar combination of glycerin and sucrose – the main carbohydrate in molasses) is converted to acetic acid (HAc), propionic acid (HPr), butyric acid (HBu) and valeric acid (HVa).

Hydrolysis and Acidogenesis

$$C_{12}H_{22}O_{11} + 2\ C_3H_8O_3 + 5\ H_2O \overset{k_1}{\rightarrow} 7\ CH_3COOH + 4\ CO_2 + 10\ H_2 \tag{1}$$

$$C_{12}H_{22}O_{11} + C_3H_8O_3 + 4\ H_2 \overset{k_2}{\rightarrow} 5\ CH_3CH_2COOH + 4\ H_2O \tag{2}$$

$$C_{12}H_{22}O_{11} + 3\ C_3H_8O_3 + 4\ H_2O \overset{k_3}{\rightarrow} 3\ CH_3CH_2CH_2COOH + 9\ CO_2 + 15\ H_2 \quad (3)$$

$$C_{12}H_{22}O_{11} + 2C_3H_8O_3 + 3H_2O \overset{k_4}{\rightarrow} 2\ CH_3CH_2CH_2CH_2COOH + 8\ CO_2 + 12\ H_2 \quad (4)$$

In the next three in series steps (*acetogenesis*), propionic, butyric and valeric acids are consumed to form smaller chain acids and hydrogen (H).

Acetogenesis

$$CH_3CH_2COOH + 2\ H_2O \overset{k_6}{\rightarrow} CH_3COOH + CO_2 + 3\ H_2 \quad (5)$$

$$CH_3CH_2CH_2COOH + 2\ H_2O \overset{k_7}{\rightarrow} CH_3CH_2COOH + CO_2 + 3\ H_2 \quad (6)$$

$$CH_3CH_2CH_2CH_2COOH + 2\ H_2O \overset{k_8}{\rightarrow} CH_3CH_2CH_2COOH + CO_2 + 3\ H_2 \quad (7)$$

In the next two independent steps (*methanogenesis*), there are the aceticlastic and hydrogenotrophic routes produce methane (M).

Methanogenesis

$$CH_3COOH \overset{k_9}{\rightarrow} CH_4 + CO_2 \quad (8)$$

$$4\ H_2 + CO_2 \overset{k_{10}}{\longrightarrow} CH_4 + 2\ H_2O \quad (9)$$

Equations 10 to 16 present the first order reaction rate equations.

Equations 10 to 16 present the first order reaction rate equations of substrate consumption (r_S), formation and consumption of acetic acid (r_{HAc}), propionic acid (r_{HPr}), butyric acid (r_{HBu}), valeric acid (r_{HVa}), hydrogen (r_H) and methane formation (r_M), respectively. The kinetic parameter "k" is related to the reaction rate, indicating a relation to the time that is necessary for the concentration (S, HAc, HPr, HBu, HVa, H or M) to reach a residual value in accordance with the hypothesis of the kinetic model. Subscripts "1, 2, 3, 4, 5, 6, 7, 8 and 9" are related to the reactions. Subscripts "S, HAc, HPr, HBu, HVa, H e M" are related to the experimental values used to calculate the parameters.

$$r_S = -(k_{1S} + k_{2S} + k_{3S} + k_{4S}) \cdot C_S = -k'_{1S} \cdot C_S \tag{10}$$

$$r_{HAc} = k_{1HAc} \cdot C_S + k_{7HAc} \cdot C_{HPr} - k_{8HAc} \cdot C_{HAc} \tag{11}$$

$$r_{HPr} = k_{2HPr} \cdot C_S + k_{6HPr} \cdot C_{HBu} - k_{7HPr} \cdot C_{HPr} \tag{12}$$

$$r_{HBu} = k_{3HBu} \cdot C_S + k_{5HBu} \cdot C_{HVa} - k_{6HBu} \cdot C_{HBu} \tag{13}$$

$$r_{HVa} = k_{4HVa} \cdot C_S - k_{5HVa} \cdot C_{HVa} \tag{14}$$

$$r_H = k_{1H} \cdot C_S - k_{2H} \cdot C_S + k_{3H} \cdot C_S + k_{4H} \cdot C_S + k_{5H} \cdot C_{HVa} + k_{6H} \cdot C_{HBu} + k_{7H} \cdot C_{HPr} - k_{9H} \cdot C_H \tag{15}$$

$$r_M = k_{8M} \cdot C_{HAc} + k_{9M} \cdot C_H \tag{16}$$

Equations 17 to 24 present the mass balance of the reactor in fed-batch mode concerning the kinetic model (substrate, volatile acids and methane). These equations were used to determine the kinetic parameters of the model. Subscripts "INF" are related to the concentrations of compounds in the influent.

$$\frac{dV}{dt} = F \tag{17}$$

$$\frac{dC_S}{dt} = \frac{F}{V} \cdot (C_{S\,INF} - C_{SF}) + r_S \tag{18}$$

$$\frac{dC_{HAc}}{dt} = \frac{F}{V} \cdot (C_{HAc\,INF} - C_{HAc}) + r_{HAc} \tag{19}$$

$$\frac{dC_{HPr}}{dt} = \frac{F}{V} \cdot (C_{HPr\,INF} - C_{HPr}) + r_{HPr} \tag{20}$$

$$\frac{dC_{HBu}}{dt} = \frac{F}{V} \cdot (C_{HBu\,INF} - C_{HBu}) + r_{HBu} \tag{21}$$

$$\frac{dC_{HVa}}{dt} = \frac{F}{V} \cdot (C_{HVa\,INF} - C_{HVa}) + r_{HVa} \tag{22}$$

$$\frac{dC_{H}}{dt} = \frac{F}{V} \cdot (C_{H\,INF} - C_{H}) + r_{H} \tag{23}$$

$$\frac{dC_{M}}{dt} = -\frac{F}{V} \cdot (C_{M}) + r_{M} \tag{24}$$

To address the differential equations, the Euler numerical integration method (Excel software) was used. These parameters were calculated using an objective function in the optimization procedure (function Solver in Excel software) for the minimum square error between experimental and model data.

2.7. Estimated Scale-Up and Energy Production

The method used to estimate the reactor on an industrial scale and energy production was based Albanez et al. (2016). The glycerin volume to be treated per day was calculated using biodiesel production capacity values of small and medium-sized industries. To estimate the energy produced by the industrial scale reactor, biogas production (methane) and the potential energy released by methane combustion was used as a design parameter. Data from assay CV was used in the estimative, since this assay achieved the best performance indicators in relation to methane production.

2.8. Experimental Procedure

Table 1 presents the operational conditions implemented in the AnSBBR, which were separated in the following phases: Preliminary, Adaptation, Mono-digestion, Co-digestion and Feeding Strategy. The AnSBBR was operated in batch (B) and fed batch (FB) mode with influent concentration (C_S) from 1000 to 7000 mgCOD.L^{-1}, 8 h of cycle length, temperature from 30 to 55°C and stirring frequency from 50 to 100 rpm.

Table 1. Summary of the investigated conditions

Phases	Assay	Operation	C_S (gCOD.L^{-1})	T (°C)	Duration (d)	%G/%M
Preliminary	PI	B	1.0	30	2	100/0
			2.0	35	2	
			3.0	40/45	3	
			4.0	50	2	
	PII	B	5.0	55	11	
Adaption	AI	B	2.5	55	3	50/50
	AII	B	2.0		21	
Mono-digestion	MI	B	2.0	55	21	100/0
	MII	B	3.0		21	
	MIII	B	4.0		21	
	MIV	B	5.0		21	
	MV	B	6.0		21	
Co-digestion	CI	B	3.0	55	17	66/33
	CII	B	4.0		20	75/25
	CIII	B	5.0		20	80/20
	CIV	B	6.0		22	83/17
	CV	B	7.0		20	86/14
Feeding Strategy	CV-BA	BA	7.0	55	12	86/14

G: Glycerin, M: Molasses.

In the Preliminary phase, influent was made with 100% glycerin and its concentration varied from 1000 to 5000 mgCOD.L^{-1}. In the Adaptation phase, influent composition was 50% glycerin and 50% molasses and its concentration was 2500 and 2000 mgCOD.L^{-1}. In the Mono-digestion phase, influent was made with 100% glycerin and its concentration was increased from 2000 to 6000 mgCOD.L^{-1}. In the Co-digestion stage, molasses concentration was constant (1000 mgCOD.L^{-1}) and influent concentration varied from 3000 to 7000 mgCOD.L^{-1}. Finally, in the Feeding Strategy phase, the same conditions from Assay CV were applied to the system but using the fed batch mode.

In the assays operated in batch mode (B), the filling time was equal to 10 minutes, which is 2% of the total cycle length. In the fed-batch operation (FB), filling time was 4 h (50% of cycle length). In all assays, influent was supplemented with urea (CH_4N_2O), due to its the nitrogen source, with a ratio of 0.0058 mgCH_4N_2O.mgCOD^{-1}.

The Preliminary phase was carried out aiming at microorganism's adaptation, by means of the gradual increase in influent concentration (1000 to 5000 mgCOD.L^{-1}) and in temperature (30 to 55°C). Initially, the basket containing the immobilized biomass was allocated inside the reactor, and immobilized biomass was also allocated under the basket. Next, in the assay PI, 3.9 L of influent were fed to the system with 1.0 g$NaHCO_3$.gDQO^{-1}. The ultra-thermostatic bath was turned on at 30°C with 100 rpm of stirring speed.

After 8h of this cycle, stirring was interrupted and discharge began. 1.0 L of effluent was discharged in 10 min, maintain 1.8 L of residual volume inside the reactor. After the reactor discharge, a time interval of 1 min was allowed as a safety period for synchronizing the operation of the timer-controlled feed and discharge pumps, and then the next cycle was started. At the beginning of a new cycle, stirring was again turned on along with the feeding of 1.0 L of influent. Thus, the cycle was repeated, characterizing the sequencing batches of the AnSBBR.

Then, still in PI assay, the start-up strategy was to increase influent concentration in 1000 mgCOD.L^{-1} and temperature in 5°C every 2 days for the first two weeks. In the third week, concentration was maintained at 5000 mgCOD.L^{-1} and the temperature at 55°C (PII assay).

Adaptation phase was performed aiming to favor growth of methanogenic microorganism, by digesting molasses and glycerin (50:50) with an influent concentration of 2000 mgCOD.L^{-1}. Molasses was chosen as a co-substrate due to its biodegradability characteristics, since they favor biomass adaptation and development of methanogenic microorganisms (Misi and Forster 2001). In the first two days, substrate concentration was 2500 mgCOD.L^{-1} (AI Assay) and 2000 mgCOD.L^{-1} in the following days.

$NaHCO_3.DQO^{-1}$ ratio was reduced to 0.8 and stirring was reduced to 50 rpm. Temperature was maintained at 55°C as well as cycle time in 8h and feeding of 1.0 L of influent in 10 min.

The Mono-digestion phase was carried out aiming to verify the influence of the increase in influent concentration, by the gradual increase of glycerin concentration, from 2000 to 6000 $mgCOD.L^{-1}$.

In Co-digestion phase, influent concentration varied from 3000 to 7000 $mgCOD.L^{-1}$. Molasses concentration was constant (1000 $mgCOD.L^{-1}$). Consequently, influent composition was variable, being 66%/33%; 75%/25%; 80%/20%; 83%/17% e 86%/14% of glycerin/molasses, respectively, in assays CI, CII, CIII, CIV e CV (Table 1). Co-digestion was performed to verify it there would be improvement in process performance with the addition of a substrate easier to degrade. $NaHCO_3.DQO^{-1}$ ratio was 0.8 in Mono-digestion phase and 1.0 in Co-digestion phase. Temperature was maintained at 55°C as well as the feeding of 1.0 L of influent in 10 min.

In the Feeding strategy phase, the influence of filling time was evaluated. The AnSBBR was operated under the same conditions as assay CV, but its feeding was performed in 4 h, with a cycle length of 8h. Bicarbonate/COD ratio was 0.1, temperature was 55°C and stirring frequency was 50 rpm.

After reaching stability, profiles were taken of the following: soluble organic matter, carbohydrates and glycerin concentrations, bicarbonate alkalinity, total volatile acids, intermediate metabolites (acetone, volatile acids and alcohols), pH and biogas (composition and production). These profiles allowed a better understanding of the metabolic routes during a cycle. To obtain these profiles, samples were taken at 30 to 60-min time intervals. The volume that was collected did not exceed 300 mL and was always less than 10% of the total volume of wastewater in the reactor. Thus, a new experimental condition was implemented by changing the influent concentration and/or cycle length and feeding time.

3. RESULTS AND DISCUSSION

3.1. Mono-Digestion

In mono-digestion phase, influent was made with bi-distilled glycerin. The AnSBBR was operated in batch mode, with 8h of cycle length, 10 min of feeding time and at 55°C. Five assays were carried out (MI to MV) with increasing organic loading rate: 2.3 $kgCOD.m^{-3}.d^{-1}$ (MI), 3.2 $kgCOD.m^{-3}.d^{-1}$ (MII), 4.3 $kgCOD.m^{-3}.d^{-1}$ (MIII), 5.5 $kgCOD.m^{-3}.d^{-1}$ (MIV) e 6.5 $kgCOD.m^{-3}.d^{-1}$ (MV). Table 2 shows the results achieved in this phase.

The AnSBBR was stable in all assays. Organic matter was consumed and intermediate volatile acids were generated and consumed with methane and carbon dioxide production. The increase in the OLR_A decreased organic matter removal efficiency (ε_{SF}) from 77 ± 2% (MI) to 55 ± 4% (MV), however glycerin removal efficiency (ε_{GF}) was close to 100% in all assays.

There was also an increase in total volatile fatty acids (TVFA) in effluent samples, from 257 $mgHAc.L^{-1}$ (MI) to 2229 $mgHAc.L^{-1}$ (MV). Consequently, there was a higher consumption of bicarbonate alkalinity (BA) (from 204 to 1982 $mgCaCO_3.L^{-1}$) and pH fell from 7.93 (MI) to 7.27 (MV), without compromising process stability.

Total volatile solids mass (M_{SVT}) in the reactor was higher in the assays operated with higher OLR_A (MIV and MV). There was not significant biomass loss in any of the assays, which means that the system was not submitted to critical enough conditions that would cause biomass wash-out.

The intermediate volatile acids concentration (acetic, propionic, butyric and valeric acids) also increased with higher OLR_A. For the lowest OLR_A (MI), there was a predominance of propionic acid, while in the other assays (MII and MV) there was more acetic acid.

Methane production also increased with increased influent concentration. As a result, the highest methane molar productivity (MPr) – 64.4 $molCH_4.m^{-3}.d^{-1}$ - was achieved with the highest OLR_A (MV).

Regarding the yield between methane produced and organic matter applied (Y_A), similar values of this value were obtained until 4.3 $gCOD.L^{-1}.d^{-1}$ (MIII), and then there was a decrease in this parameter. Therefore, the increase in the organic loading rate did not cause an improvement in methane molar yield, since the methane production gain was not enough to overcome the increase in the applied organic load. The best yield was achieved in assay MII (12.3 $molCH_4.kgCOD^{-1}$). Biogas quality was similar in all assays, with methane percentage varying among 64 to 69%.

3.2. Co-Digestion

In co-digestion phase, influent was made with bi-distilled glycerin and molasses (co-substrate). Molasses was added to the reaction medium with the objective of improving organic matter removal when compared to mono-digestion due to its better biodegradability.

The AnSBBR was operated in batch and fed-batch mode with 8h of cycle length. In batch operation, feeding time was equal to 10 min and in fed-batch operation, it was equal to 4h. Six assays were carried out (CI to CV and CV-BA) with increasing organic loading rate: 3.2 $kgCOD.m^{-3}.d^{-1}$ (CI), 4.6 $kgCOD.m^{-3}.d^{-1}$ (CII), 5.6 $kgCOD.m^{-3}.d^{-1}$ (CIII), 6.7 $kgCOD.m^{-3}.d^{-1}$ (CIV), 7.7 $kgCOD.m^{-3}.d^{-1}$ (CV) e 7.5 $kgCOD.m^{-3}.d^{-1}$ (CV-FB). Table 3 shows the results achieved in this phase.

The AnSBBR operation was stable in all assays. For OLR_A up to 4.6 $kgCOD.m^{-3}.d^{-1}$, organic matter removal efficiency was 93% and then it decreased with higher organic loads. This decrease could be related to the reduction in molasses percentage in the influent. Glycerin/molasses percentages were 66%/33% (CI); 75%/25% (CII); 80%/20% (CIV) and 83%/17% (CV). Therefore, considering the obtained results, the best mixture percentage was 75% glycerin with 25% molasses.

Glycerin removal efficiency was close to 100% in all assays and carbohydrates removal efficiency was constant in 97 ± 1% even with the increase in the OLR_A.

There was an increase in total volatile fatty acids concentration in effluent samples (from 24 to 1997 $mgHAc.L^{-1}$) with higher OLR_A. As a result, bicarbonate alkalinity was increasingly consumed (from 7 to 1709 $mgCaCO_3.L^{-1}$) but with constant pH. In assay CI, exceptionally, there was a generation of alkalinity in relation to influent alkalinity (52 $mgCaCO_3.L^{-1}$).

Volatile intermediate acids concentration also increased with higher OLR_A. There was a prevalence of acetic acid in all assays.

As in mono-digestion, the increase in the organic loading rate in co-digestion also improved methane molar productivity. The best value of this parameter was 84.4 $molCH_4.m^{-3}.d^{-1}$ and it was achieved with the highest OLR_A (CV).

Regarding the methane molar yield per applied organic loading rate, the highest value of this parameter (14.2 $molCH_4.kgCOD^{-1}$) was achieved with the smaller OLR_A. This parameter decreased with higher organic loading rates.

Nevertheless, the methane molar yield per removed organic loading rate increased with the increase in OLR_A. The same tendency is observed with methane specific molar productivity (SMPr).

Biogas quality was similar in all assays: from 66 to 72% of methane.

Regarding operation in fed-batch mode (CV-FB), organic matter removal (59 ± 3%) was similar to operation in batch (63 ± 1%). Glycerin and carbohydrate removal efficiencies were not altered by the feeding mode.

The fed-batch mode did not cause total volatile acids accumulation or a greater consumption of alkalinity, as seen in Table 3.

All performance indicators were close when comparing the both strategies of feeding. These results show that the system is flexible and it can be operated in both modes for the co-digestion of glycerin and molasses in thermophilic conditions.

Table 2. Mean values of the parameters and performance indicators of the reactor at Mono-digestion

Parameter		MI				MII				MIII				MIV				MV			
ε_{SF}	(%)	77	±	2	(14)	74	±	3	(9)	66	±	2	(10)	60	±	3	(10)	55	±	4	(11)
ε_{GF}	(%)	100	±	0	(15)	100	±	0	(11)	100	±	0	(13)	99	±	1	(12)	100	±	0	(10)
M_{TVS}	(g)	14.2				14.2				14.2				16.6				21.1			
OLR_A	($kgCOD.m^{-3}.d^{-1}$)	2.3				3.2				4.3				5.5				6.5			
OLR_R	($kgCOD.m^{-3}.d^{-1}$)	1.8				2.4				2.9				3.3				3.6			
Influent	pH (u)	8.21	±	0.08	(15)	8.48	±	0.33	(14)	8.46	±	0.08	(14)	8.48	±	0.06	(14)	8.56	±	0.07	(12)
	TA ($mgCaCO_3.L^{-1}$)	1031	±	44	(13)	1605	±	88	(14)	1964	±	38	(10)	2463	±	51	(11)	2893	±	49	(9)
	BA ($mgCaCO_3.L^{-1}$)	1010	±	45	(13)	1580	±	88	(14)	1927	±	40	(10)	2433	±	52	(11)	2853	±	50	(9)
	TVFA ($mgHAc.L^{-1}$)	29	±	10	(13)	35	±	6	(14)	52	±	12	(10)	43	±	7	(11)	56	±	7	(9)
Effluent	pH (u)	7.93	±	0.24	(15)	8.16	±	0.27	(13)	7.37	±	0.17	(14)	7.27	±	0.10	(14)	7.27	±	0.07	(11)
	TA ($mgCaCO_3.L^{-1}$)	988	±	26	(13)	1507	±	22	(10)	1779	±	36	(10)	2178	±	73	(11)	2453	±	50	(8)
	BA ($mgCaCO_3.L^{-1}$)	806	±	86	(13)	1162	±	112	(10)	1014	±	96	(10)	1005	±	93	(11)	871	±	78	(8)
	TVFA ($mgHAc.L^{-1}$)	257	±	106	(13)	486	±	142	(10)	1078	±	126	(10)	1652	±	75	(11)	2229	±	110	(8)
VFA (molar %)	Acetic acid	36.5				54.0				54.9				54.7				60.0			
	Propionic acid	57.1				36.5				35.5				36.6				32.0			
	Butyric acid	6.4				8.3				8.2				7.4				6.8			
	Valeric acid	0.0				1.2				1.4				1.3				1.1			
Total concentration ($mmol.L^{-1}$)		4.1				7.2				11.1				32.1				44.0			
MPr	($molCH_4.m^{-3}.d^{-1}$)	27.2				39.6				50.9				56.8				64.4			
SMPr	($molCH_4.kgSVT^{-1}.d^{-1}$)	5.5				8.1				10.2				9.6				8.3			
Y_A	($molCH_4.kgDQO^{-1}$)	11.8				12.3				11.8				10.3				10.0			
Y_R	($molCH_4.kgDQO^{-1}$)	15.3				16.7				17.9				17.2				18.0			
Gas	(%) CH_4	65.0				68.8				66.9				64.3				65.7			
	(%) CO_2	35.0				31.2				33.1				35.7				34.3			

Table 3. Mean values of the parameters and performance indicators of the reactor at Co-digestion

Parameter		AII	CI	CII	CIII	CIV	CV	CV-BA
ε_{SF}	(%)	76 ± 6 (12)	94 ± 0 (10)	93 ± 1 (12)	81 ± 3 (11)	71 ± 2 (11)	63 ± 1 (11)	59 ± 3 (6)
ε_{GF}	(%)	100 ± 0 (9)	100 ± 0 (9)	100 ± 0 (14)	100 ± 0 (11)	100 ± 0 (13)	100 ± 0 (13)	100 ± 0 (8)
ε_{CF}	(%)	98 ± 0 (13)	97 ± 0 (10)	97 ± 0 (12)	97 ± 0 (10)	96 ± 1 (9)	96 ± 1 (11)	95 ± 1 (8)
M_{TVS}	(g)	14.2	21.1	28.3	28.3	28.3	28.3	49.5
OLR_A	($kgCOD.m^{-3}.d^{-1}$)	2.4	3.2	4.6	5.6	6.7	7.7	7.5
OLR_R	($kgCOD.m^{-3}.d^{-1}$)	1.8	3.0	4.3	4.6	4.7	4.9	4.4
Influent	pH (u)	8.42 ± 0.28 (14)	8.46 ± 0.06 (10)	8.47 ± 0.07 (12)	8.41 ± 0.11 (12)	8.51 ± 0.06 (14)	8.50 ± 0.06 13	8.52 ± 0.06 (8)
	TA ($mgCaCO_3.L^{-1}$)	1089 ± 26 (12)	1798 ± 29 (7)	1203 ± 70 (12)	2814 ± 106 (9)	3480 ± 103 (11)	4018 ± 115 (12)	4085 ± 168 (7)
	BA ($mgCaCO_3.L^{-1}$)	1048 ± 26 (12)	1758 ± 33 (7)	2366 ± 70 (12)	2767 ± 99 (9)	3435 ± 99 (11)	3962 ± 113 (12)	4033 ± 161 (7)
	TVFA ($mgHAc.L^{-1}$)	57 ± 22 (12)	56 ± 18 (7)	57 ± 6 (12)	67 ± 18 (9)	64 ± 12 (11)	79 ± 22 (12)	73 ± 12 (7)
Effluent	pH (u)	8.12 ± 0.15 (14)	7.51 ± 0.14 (10)	7.62 ± 0.09 (11)	7.60 ± 0.06 (11)	7.64 ± 0.10 (13)	7.62 ± 0.05 (13)	7.61 ± 0.11 (9)
	TA ($mgCaCO_3.L^{-1}$)	1057 ± 45 (12)	1827 ± 25 (7)	2411 ± 58 (11)	2717 ± 98 (8)	3225 ± 51 (10)	3671 ± 109 (12)	3619 ± 183 (6)
	BA ($mgCaCO_3.L^{-1}$)	803 ± 147 (12)	1810 ± 24 (7)	2359 ± 54 (11)	2269 ± 197 (8)	2272 ± 88 (10)	2253 ± 128 (12)	2255 ± 194 (6)
	TVFA ($mgHAc.L^{-1}$)	358 ± 166 (12)	24 ± 2 (7)	73 ± 9 (11)	631 ± 141 (8)	1342 ± 106 (10)	1997 ± 89 (12)	1922 ± 65 (6)
VFA (molar %)	Acetic acid	77.4	59.4	58.3	59.5	66.1	69.6	66.9
	Propionic acid	18.7	32.6	32.5	31.1	25.8	24.9	27.2
	Butyric acid	3.9	6.6	6.2	6.3	5.4	3.5	4.0
	Valeric acid	0.0	1.4	3.0	3.1	2.7	2.0	2.0
Total concentration ($mmol.L^{-1}$)		4.7	4.4	5.0	7.5	13.1	23.3	17.5
MPr	($molCH_4.m^{-3}.d^{-1}$)	28.3	44.7	60.3	70.5	72.8	84.4	83.7
SMPr	($molCH_4.kgSVT^{-1}.d^{-1}$)	5.7	5.9	6.0	7.1	7.4	8.5	5.1
Y_A	($molCH_4.kgDQO^{-1}$)	11.7	14.2	13.0	12.5	10.9	11.0	11.2
Y_R	($molCH_4.kgDQO^{-1}$)	15.4	15.0	14.0	15.4	15.3	17.3	18.9
Gas	(%) CH_4	56.4	65.8	67.8	70.8	71.2	71.6	69.1
	(%) CO_2	43.6	34.2	32.2	29.2	28.8	28.4	30.9

Nevertheless, the methane molar yield per removed organic loading rate increased with the increase in OLR_A. The same tendency is observed with methane specific molar productivity (SMPr).

Biogas quality was similar in all assays: from 66 to 72% of methane.

Regarding operation in fed-batch mode (CV-FB), organic matter removal (59 ± 3%) was similar to operation in batch (63 ± 1%). Glycerin and carbohydrate removal efficiencies were not altered by the feeding mode.

The fed-batch mode did not cause total volatile acids accumulation or a greater consumption of alkalinity, as seen in Table 3.

All performance indicators were close when comparing the both strategies of feeding. These results show that the system is flexible and it can be operated in both modes for the co-digestion of glycerin and molasses in thermophilic conditions.

3.3. Comparing Mono-Digestion and Co-Digestion

Figures 2 and 3 show organic matter removal efficiency, total volatile fatty acids concentration, methane productivity and yield in mono-digestion (M), co-digestion operated in batch (C) and co-digestion operated in fed-batch (CV-FB) assays.

The increase in the organic loading rate caused organic matter removal efficiency to decrease both in mono-digestion (from 77 to 55%) and co-digestion (from 94 to 59%) assays. Nevertheless, co-digestion achieved higher removal efficiencies, probable due to better nutrient balance, increase in the biodegradable organic load and lower concentration of TFVA, which inhibit methanogenic activity. Accordingly, the removed organic loading rate (OLR_R) was higher in co-digestion assays; it achieved from 3.0 $kgCOD.m^{-3}.d^{-1}$ (CI) to 4.9 $kgCOD.m^{-3}.d^{-1}$ (CV), while in mono-digestion, the same parameter varied from 1.8 $kgCOD.m^{-3}.d^{-1}$ (MI) to 3.6 .$kgCOD.m^{-3}.d^{-1}$ (MV).

Regarding this last advantage of the co-digestion, TFVA concentration was always lower in co-digestion assays, varying from 24 ± 2 $mgHAc.L^{-1}$ (CI) to 1997 ± 89 $mgHAc.L^{-1}$ (CV). In mono-digestion assays, the same parameter varied from 257 ± 106 mgHAc.L^{-1} (MI) to 2229 ± 110 mgHAc.L^{-1} (MV). It is noteworthy that assay MV (6.5 kgCOD.m^{-3}.d^{-1}) was operated under a smaller OLR_A than assay CV (7.7 kgCOD.m^{-3}.d^{-1}), which shows the advantage of using a co-digestion process due to its better equilibrium between generation and consumption of intermediate compounds.

Comparing all intermediate compounds, acetic acid achieved the highest concentration both in co-digestion (3.7 to 11.7 mmolHAc.L^{-1}) and in mono-digestion assays (1.5 a 26.4 mmolHac.L^{-1}), which is most favorable for methane production (Lovato et al., 2018). As for propionic acid concentration, it increased with the increment in organic loading rate. In mono-digestion assays, its concentration went from 0.17 gHPr.L^{-1} (MI) to 1.04 gHPr.L^{-1} (MV). In co-digestion assays, HPr concentration varied between 0.43 gHPr.L^{-1} (CV-B) and 0.35 gHPr.L^{-1} (CV-FB). Therefore, there was not an impairment for methane production, since methanogenic *archaea* are inhibited in concentrations higher than 1-2 g.L^{-1}, while the tolerate acetic and butyric concentrations up to 10 g.L^{-1} (Lovato et al., 2018).

Thus, the increase in the OLR_A hindered the total consumption of the generated organic acids in acidogenesis, consequently, the organic matter removal efficiency decreased as influent concentration increased.

In co-digestion, the process was more efficient in bicarbonate alkalinity generation during the cycle, from microbial metabolism. Bicarbonate alkalinity in effluent samples varied from 1810 ± 24 mg$CaCO_3$.L^{-1} (CI) to 2253 ± 128 mg$CaCO_3$.L^{-1} (CV), while in mono-digestion, the same parameter varied from 806 ± 86 mg$CaCO_3$.L^{-1} (MI) to 1005 ± 93 mg$CaCO_3$.L^{-1} (MV).

Methane productivity improved with the increase in OLR_A for both typed of anaerobic digestion due to a greater availability of substrate, however, the assays operated with co-digestion achieved higher productivity values due to a more favorable environment for the anaerobic process, i.e., lower concentration od total volatile fatty acids and higher bicarbonate alkalinity. The highest PrM was achieved with the highest OLR_A (7.7 kgDQO.m^{-1}.d^{-1}): 84.4 molCH_4·m^{-3}·d^{-1}. This value was very close to the one achieved in the assay operated in fed-batch mode (83,7 molCH_4·m^{-3}·d^{-1}). Hence, this parameter was not influenced significantly by the feeding mode.

Regarding the yield between produced methane and applied organic load, in mono-digestion, the highest value -12.3 molCH4.kgCOD^{-1} - was achieved in assay MII (2.3 kgCOD.m^{-3}.d^{-1}). Under higher organic loads, this parameter decreased due to a higher concentration of total volatile acids. The same tendency was observed in co-digestion assays: the best yield - 14.2 molCH_4.kgCOD^{-1} - was achieved in assay CI (3.2 kgCOD.m^{-3}.d^{-1}) and then it decreased as OLR_A increased. Nevertheless, this performance indicator was always similar or better in co-digestion assays.

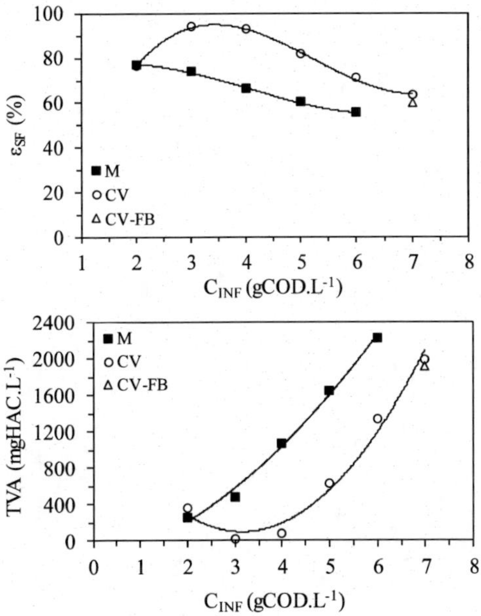

Figure 2. Organic matter removal efficiency and total volatile acids concentration in mono-digestion (M), batch operated co-digestion (C) and fed-batch operated co-digestion (CV-FB).

In mono-digestion assays, the yield between methane produced and removed organic load was slightly higher than in co-digestion assays due to a lower concentration of organic matter removed. Considering the maximum theoretical methane yield per removed load (15.6 molCH_4.kgCOD^{-1} or 350 mLCH_4.gCOD^{-1} (Heidrich, Curtis, and Dolfing, 2011), the AnSBBR achieved more than 100% of it. This is explained by the inherent experimental error in the analysis and by the consumption of particulate material that was already in the reactor from other conditions and that was used for methane generation. Still, these results indicate that the system achieved yields very close to 100%, which is great for process performance.

Thus, co-digestion in thermophilic conditions was more advantageous compared to mono-digestion. Regarding the feeding mode used, the proposed co-digestion can be operated both in batch and in fed-batch mode.

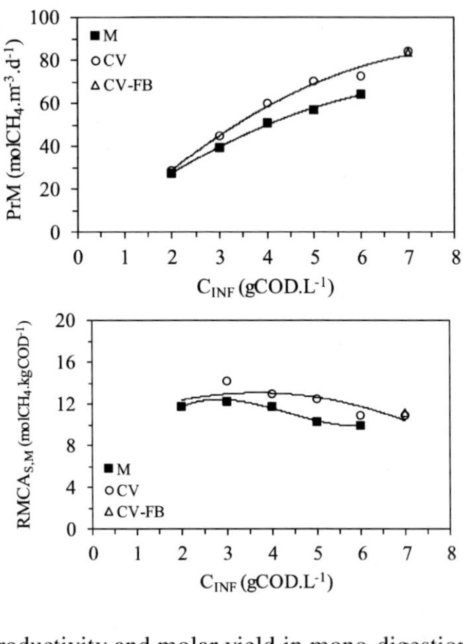

Figure 3. Methane productivity and molar yield in mono-digestion (M), batch operated co-digestion (C) and fed-batch operated co-digestion (CV-FB).

3.4. Kinetic Metabolic Model

Figure 4 shows the experimentally obtained values (markers) and the values obtained by the adjusted kinetic model (lines), both along the cycle for the main monitored variables related to the understanding of the methanogenic metabolism in the assays operated in batch (assay CV) and fed-batch (CV-FB). The fed-batch behavior of the system can be detected in the substrate concentration graphs – the substrate increases gradually up to half of the cycle length (which accounts for the feeding period) and is then consumed. Table 4 shows the values of the adjusted kinetic parameters for all assays. The model was effective in fitting the experimental data, validating interpretation of the behavior of the kinetic parameters in all experimental assays.

In glycerin mono-digestion, acetic acid (HAc) production was constant until 4000 mgCOD.L^{-1} (k_{1HAc} = 0.28 h^{-1} (MI); k_{1HAc} = 0.25 h^{-1} (MII) e k_{1HAc} = 0.26 h^{-1} (MIII)), and then increased as influent concentration increased (k_{1HAc} = 0.79 h^{-1} - 5000 mgCOD.L^{-1} and k_{1HAc} = 0.83 h^{-1} - 6000 mgCOD.L^{-1}). This behavior is also found in valeric acid (HVa) production: assays MI to MV: k_{4HVa} = 0.00 h^{-1}; k_{4HVa} = 0.05 h^{-1}, k_{4HVa} = 0.05 h^{-1}, k_{4HVa} = 0.11 h^{-1} and k_{4HVa} = 0.09 h^{-1}, respectively.

The same pattern is observed analyzing acetic acid production in co-digestion assays (Assays CI to CV-FB) (Table 4): constant production until 4000 mgCOD.L^{-1} (k_{1HAc} = 0.26 h^{-1} (CI); k_{1HAc} = 0.18 h^{-1} (CII)) and then increase with increasing influent concentration (k_{1HAc} = 0.41 h^{-1} - 5000 mgCOD.L^{-1}, k_{1HAc} = 0.76 h^{-1} - 6000 mgCOD.L^{-1} and k_{1HAc} = 1.18 h^{-1} - 7000 mgCOD.L^{-1}). Fed-batch operation also increased acetic acid production: k_{1HAc} = 11.56 h^{-1}. Regarding valeric acid production in co-digestion assays, it was, approximately, constant: k_{4HVa} = 0.01 h^{-1} (CI); k_{4HVa} = 0.02 h^{-1} (CII); k_{4HVa} = 0.02 h^{-1} (CIII); k_{4HVa} = 0.04 h^{-1} (CIV); k_{4HVa} = 0.02 h^{-1}, (CV) e k_{4HVa} = 0.03 h^{-1} (CV-FB).

Propionic acid production did not show any tendency in mono-digestion assays. In co-digestion assays, this acid production tendency was the same as acetic acid production: constant until 6000 mgCOD.L^{-1}, and then increased as influent concentration was increased and with fed-batch operation.

In mono-digestion, from influent concentration of 3000 mgCOD.L^{-1} (MII), and in co-digestion, hydrogen (H_2) is produced similarly from HAc, HBut e HVa routes (very close values of k_{1H2}, k_{3H2} e k_{4H2}).

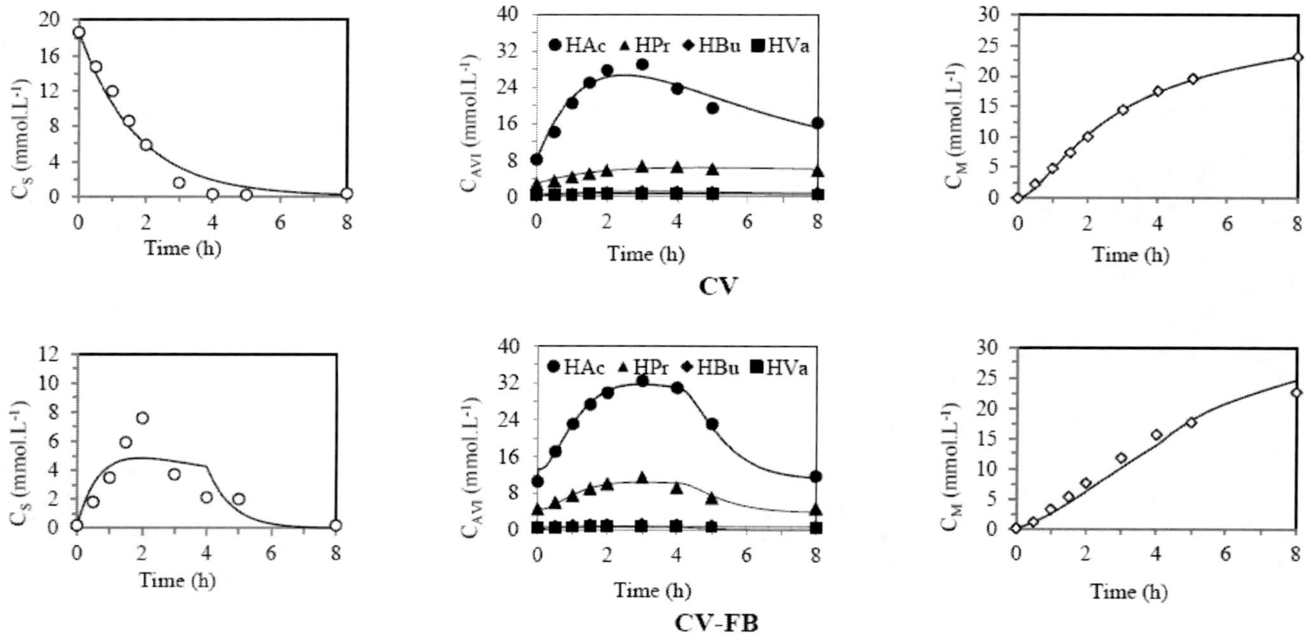

Figure 4. Substrate consumption, intermediate compounds and methane production profiles at Condition CV and CV-FB.

Table 4. Summary of the estimated kinetic parameters for all assays

Condition	Kinetic Parameter	MI	MII	MIII	MIV	MV	CI	CII	CIII	CIV	CV	CV-FB
Hydrolosis and acidogenesis	k'_{1S} (h^{-1})	0.46	0.41	0.47	0.45	0.40	0.62	0.55	0.68	0.59	0.57	1.28
	k_{1HAc} (h^{-1})	0.28	0.25	0.26	0.79	0.83	0.26	0.18	0.41	0.76	1.18	11.56
	k_{1H2} (h^{-1})	0.71	0.36	0.40	0.09	0.08	0.52	0.54	0.30	0.35	0.25	0.07
	k_{2HPr} (h^{-1})	0.31	0.12	0.13	0.32	0.33	0.09	0.07	0.10	0.10	0.14	2.54
	k_{2H2} (h^{-1})	0.71	0.00	0.00	0.00	0.00	0.12	0.12	0.20	0.25	0.16	0.09
	k_{3HBu} (h^{-1})	0.10	0.46	0.35	0.43	0.38	0.11	0.46	0.25	0.24	0.27	0.20
	k_{3H2} (h^{-1})	0.00	0.36	0.40	0.09	0.08	0.59	0.56	0.40	0.50	0.32	0.06
	k_{4HVa} (h^{-1})	0.00	0.05	0.05	0.11	0.09	0.01	0.02	0.02	0.04	0.02	0.03
	k_{4H2} (h^{-1})	0.00	0.29	0.34	0.09	0.08	0.52	0.53	0.30	0.34	0.24	0.07
Acetogenesis	k_{5HVa} (h^{-1})	0.00	0.33	0.37	0.48	0.45	0.11	0.16	0.30	0.49	0.10	0.01
	k_{5HBut} (h^{-1})	0.00	0.46	0.35	0.43	0.38	0.11	0.42	0.25	0.23	0.27	0.20
	k_{5H2} (h^{-1})	0.08	0.26	0.07	0.09	0.09	0.10	0.05	0.05	0.03	0.04	0.10
	k_{6HBut} (h^{-1})	0.00	0.22	0.12	0.21	0.23	0.03	0.14	0.01	0.03	0.12	0.00
	k_{6H2} (h^{-1})	0.00	0.00	0.05	0.10	0.10	0.09	0.04	0.00	0.12	1.43	0.10
	k_{6HPr} (h^{-1})	0.11	0.09	0.09	0.63	0.00	0.11	0.08	0.09	0.09	0.09	12.22
	k_{7HPr} (h^{-1})	0.10	0.08	0.08	0.21	0.06	0.01	0.06	0.08	0.05	0.04	1.80
	k_{7H2} (h^{-1})	4.21	0.00	0.04	0.06	0.07	0.00	0.02	0.00	0.00	0.00	0.07
	k_{7HAc} (h^{-1})	0.00	0.19	0.10	0.09	0.71	0.07	0.16	0.21	0.28	0.37	12.25
Methano-genesis	k_{8HAc} (h^{-1})	0.16	0.23	0.14	0.17	0.49	0.09	0.14	0.27	0.25	0.28	5.51
	k_{8M} (h^{-1})	0.22	0.02	0.10	0.13	0.10	0.05	0.11	0.17	0.01	0.00	0.15.
	k_{9H2} (h^{-1})	4.21	1.07	1.75	0.07	0.17	2.16	1.65	1.17	1.07	1.87	0.31
	k_{9M} (h^{-1})	1.66	0.54	0.53	0.00	0.00	1.74	0.92	0.55	0.70	1.20	0.00

Most of the acetic acid production came from acidogenesis (values of k_{1HAc} higher than k_{7HAc}), excluding condition CV-FB. Propionic acid production came similarly from acidogenesis and acetogenesis (close values of k_{2HPr} and k_{6HPr}), except assays MI and MV, in which acidogenesis is predominant, and assays MIV and CV-FB, in which acetogenesis is predominant. Butyric acid (HBu) production also came similarly from acidogenesis and acetogenesis (close values of k_{3HBu} and k_{5HBu} in all assays).

In mono-digestion, hydrogenotrophic methanogenesis (k_{9M}) was predominant until influent concentration of 4000 mgCOD.L^{-1}. Nevertheless, when the organic loading increases, methane production came prevalently from the acetoclastic route (k_{8M}).

In co-digestion, hydrogenotrophic methanogenesis was responsible for most of methane production, except in the assay operated in fed-batch mode.

3.5. Estimated Scale-Up and Energy Production

Estimation of the reactor volume on an industrial scale and its possible energy recovery were calculated based on data of Assay CV. Biodiesel production data were collected from small and medium-size industries, with production capacities of 200 and 10000 $L.d^{-1}$, respectively. To estimate the generated glycerin volume, it was considered that 0.0635 L of glycerin are generated by each liter of biodiesel produced. Hence, the daily glycerin production were 12.7 $L.d^{-1}$ and 635.0 $L.d^{-1}$, respectively, for small and medium-size industries.

The estimated working volume of the reactor for a small industry was 6.64 m^3 and 332 m^3 for a medium size industry. From these estimated volumes, a proposed configuration for the industrial operation would be of four AnSBBRs working in parallel (of 1.66 m^3 each) for a small industry and eight AnSBBRs working in parallel (of 41.6 m^3 each) for a medium-size one. Such a configuration allows continuous methane production because the process operates through parallel reactors, with successive phases of feeding, reaction, and discharging, making the reaction phase of the bio-chemical process "continuous," therefore increasing methane productivity.

Considering the biomethane production yield (17.3 $molCH_4.kgDQO^{-1}$) and methane combustion enthalpy ($\Delta H_{C-M} = 890$ $kJ.mol^{-1}$ - Perry, Green, and Maloney 1997), methane production ($Prod_M$) would be of 286 $molCH_4$ d^{-1} and 143000 $molCH_4$ d^{-1}, for small and medium-size industries, respectively. Therefore, the power generation capacity (E_M) would be of 2.94 kW and 147.1 kW.

3.6. Microbiological Test

A microbiological test was carried out with a biomass sample removed from inside the bioreactor. The predominance of microorganisms similar to methanogenic bacilli was observed; additionally, the presence of microorganisms similar to *Methanosaeta* sp (microorganisms responsible for assimilating acetate) and *Methanosarcina* sp (microorganisms respon-sible for assimilating hydrogen) was also noted, indicating a possible equili-brium of the methanogenic community.

3.7. Comparison with Literature

The reactor performance of this investigation was compared with other reactors treating glycerin as a substrate for methane production, as shown in Table 5. The comparison was performed with other reactors (ASBR or AnSBBR) operated in mesophilic conditions (30°C) and with different types of glycerin (effluent from the laboratory and industrial biodiesel production process).

Table 5. Summary of comparison with literature results

Substrate	Reactor	Operation	OLR_A	T	MPr	Y_A	ε_{SF}	Reference
EL	ASBR	B	2.44	30	14.4	5.9	73	(1)
			4.50		7.6	1.7	67	(2)
	AnSBBR	FB	7.50		21.4	2.9	77	(3)
EI	ASBR	B	2.52	30	17.0	6.7	81	(4)
		FB	3.71		15.4	4.1	81	
GB	AnSBBR	B	6.45	55	64.4	10.0	55	(5)
GB-M			6.67		72.8	10.9	71	
GB-M			7.69		84.4	11.0	63	
GB-M		FB	7.48		83.7	11.2	59	

EL: Effluent from the biodiesel production process in laboratory; EI: Effluent from the industrial biodiesel production process; GB: Bi-distilled glycerin; GB-M: Bi-distilled glycerin and molasses; B: Batch; FB: Sed batch; OLR_A ($kgCOD.m^{-3}.d^{-1}$); T (°C); MPr ($molCH_4.m^{-3}.d^{-1}$); Y_A ($molCH_4.kgCOD^{-1}$); ε_{SF}(%); (1) (Selma et al., 2010); (2) (Bezerra et al., 2011); (3) (Lovato et al., 2012); (4) (Silva et al., 2013); (5) This work.

The thermophilic reactor of the present work achieved better results regarding methane productivity and yield, both for batch operation and for fed-batch operation. The lowest values of productivity and yield achieved in this work were 27.2 $molCH_4.m^{-3}.d^{-1}$ and 11.8 $molCH_4.kgCOD^{-1}$ with 2.3 $kgCOD.m^{-3}.d^{-1}$. The highest values of these parameters achieved in other papers were 21.4 $molCH_4.m^{-3}.d^{-1}$ and 2.9 $molCH_4.kgCOD^{-1}$ with 7.5 $kgCOD.m^{-3}.d^{-1}$ (Lovato et al., 2012) and 17.0 $molCH_4.m^{-3}.d^{-1}$ and 6.7 $molCH_4.kgCOD^{-1}$ with 2.5 $kgCOD.m^{-3}.d^{-1}$ (Silva et al., 2013). Hence, for glycerin treatment, the AnSBBR operated in thermophilic conditions achieved better results and these are optimized when glycerin is co-digested with molasses.

CONCLUSION

Glycerin mono-digestion in an AnSBBR with mechanical stirring under thermophilic conditions was not effective for microorganism's adaptation. Neither the gradual increase in influent concentration (1000-5000 $mgCOD.L^{-1}$) or in temperature (30-55°C) were sufficient for achieving good process performance. Hence, it was necessary to add a co-substrate with a better biodegradability, molasses, with a concentration of 2000 $mgCOD.L^{-1}$ and temperature of 55°C.

Increasing the applied volumetric organic loading rate decreased organic matter removal efficiency in mono-digestion (from 77 to 55%) and in co-digestion (from 94 to 59%), which was due to volatile fatty acids accumulation (less pronounced in co-digestion). Methane productivity increased with higher OLR for both cases, since there was more substrate available. Co-digestion (glycerin/molasses) under thermophilic conditions (55°C) achieved a better performance compared to mono-digestion (glycerin). It achieved a productivity of 84.4 $molCH4.m^{-3}.d^{-1}$ and a yield of 18.9 $molCH_4.kgDQO^{-1}$ with an OLR_A of 7.7 $gDQO.L^{-1}.d^{-1}$. The feeding strategy (batch or fed-batch) did not have a significant effect on process performance.

Analyzing the kinetic parameters in mono-digestion, methane production comes predominately from the hydrogenotrophic route until an OLR_A

of 4.3 gCOD.L^{-1}.d^{-1}. Nevertheless, with the increase in the organic load, aceticlastic methanogenesis became the dominant route. In the co-digestion phase operated in batch mode, hydrogenotrophic methanogenesis was dominant, but in fed-batch mode, there was more methane coming from the Aceticlastic route.

The scale-up estimative for the treatment of biodiesel residues in small and medium-sized industries resulted in reactors with 6.64 e 332 m^3 of working volumes, respectively. It could be achieved 286 molCH_4.d^{-1} (small size) and 14300 molCH_4.d^{-1} (medium size) with approximately 2.94 kW (small size) and 147 kW (medium size) of energy generated by methane combustion.

ACKNOWLEDGMENTS

This study was supported by the São Paulo Research Foundation (FAPESP, #14/07.692-8 #15/06.246-7), the Coordination for the Improvement of Higher Education Personnel (CAPES) and the National Council for Scientific and Technological Development (CNPq #443181/2016-0).

REFERENCES

Albanez, R., G. Lovato, M. Zaiat, S. M. Ratusznei, and J. A D Rodrigues. 2016. "Optimization, Metabolic Pathways Modeling and Scale-up Estimative of an AnSBBR Applied to Biohydrogen Production by Co-Digestion of Vinasse and Molasses." *International Journal of Hydrogen Energy* 41 (45): 20473–20484. doi:10.1016/j.ijhydene.2016.08.145.

Albuquerque, J. N., M. R. Orellana, S. M. Ratusznei, and J. A. D. Rodrigues. 2019. "Thermophilic Biomethane Production by Vinasse in an AnSBBR: Start-up Strategy and Performance Optimization." *Brazilian Journal of Chemical Engineering*. (in press).

Almeida, W. A., S. M. Ratusznei, M. Zaiat, and J. A. D. Rodrigues. 2017. "AnSBBR Applied to Biomethane Production for Vinasse Treatment: Effects of Organic Loading, Feed Strategy and Temperature." *Brazilian Journal of Chemical Engineering* 34 (3): 759–773. doi:10.1590/0104-6632.20170343s20150584.

APHA/AWWA/WEF. 2012. "Standard Methods for the Examination of Water and Wastewater." *Standard Methods*, 541. doi:ISBN 9780875 532356.

Bezerra, Roberto Antonio, José Alberto Domingues Rodrigues, Suzana Maria Ratusznei, Catarina Simone Andrade Canto, and Marcelo Zaiat. 2011. "Effect of Organic Load on the Performance and Methane Production of an AnSBBR Treating Effluent from Biodiesel Production." *Applied Biochemistry and Biotechnology* 165 (1): 347–368. doi:10.1007/s12010-011-9255-6.

Bondioli, Paolo, and Laura Della Bella. 2005. "An Alternative Spectrophotometric Method for the Determination of Free Glycerol in Biodiesel." *European Journal of Lipid Science and Technology* 107 (3): 153–157. doi:10.1002/ejlt.200401054.

Dubois, Michel, K. A. Gilles, J. K. Hamilton, P. A. Rebers, and Fred Smith. 1956. "Colorimetric Method for Determination of Sugars and Related Substances." *Analytical Chemistry* 28 (3): 350–356. doi:10.1021/ac601 11a017.

Heidrich, E. S., T. P. Curtis, and J. Dolfing. 2011. "Determination of the Internal Chemical Energy of Wastewater." *Environmental Science and Technology* 45 (2): 827–832. doi:10.1021/es103058w.

Kurahashi, Kensuke, Chie Kimura, You Fujimoto, and Hayato Tokumoto. 2017. "Value-Adding Conversion and Volume Reduction of Sewage Sludge by Anaerobic Co-Digestion with Crude Glycerol." *Bioresource Technology* 232: 119–125. doi:10.1016/j.biortech.2017.02.012.

Lovato, G., C. Z. Lazaro, M. Zaiat, S. M. Ratusznei, and J. A. D. Rodrigues. 2017. "Biohydrogen Production by Co-Digesting Whey and Glycerin in an AnSBBR: Performance Optimization, Metabolic Pathway Kinetic Modeling and Phylogenetic Characterization." *Biochemical Engineering Journal* 128: 93–105. doi:10.1016/j.bej.2017.09.011.

Lovato, G, Roberta Albanez, D. M. F. Lima, I. S. M. Bravo, W. A. Almeida, S. M. Ratuznei, and J. A. D. Rodrigues. 2015. *Application and Environmental Compliance of Anaerobic Sequencing Batch Reactors Applied to Hydrogen/Methane Bioenergy Production. Wastewater Treatment: Processes, Management Strategies and Environmental/ Health Impacts.* 1st ed. Hauppauge: Nova Science Publishers. https://www.scopus.com/inward/record.uri?eid=2-s2.0-84956732534&partnerID=40&md5=851bf2743af17dc471febba40b7e96e2.

Lovato, Giovanna, Roberta Albanez, Marianne Triveloni, Suzana M Ratusznei, and José A D Rodrigues. 2018. "Methane Production by Co-Digesting Vinasse and Whey in an AnSBBR: Effect of Mixture Ratio and Feed Strategy." *Applied Biochemistry and Biotechnology*, 1–19. doi:10.1007/s12010-018-2802-7.

Lovato, Giovanna, Roberto Antonio Bezerra, José Alberto Domingues Rodrigues, Suzana Maria Ratusznei, and Marcelo Zaiat. 2012. "Effect of Feed Strategy on Methane Production and Performance of an AnSBBR Treating Effluent from Biodiesel Production." *Applied Biochemistry and Biotechnology* 166 (8): 2007–2029. doi:10.1007/s12010-012-9627-6.

Maragkaki, A. E., M. Fountoulakis, A. Gypakis, A. Kyriakou, K. Lasaridi, and T. Manios. 2017. "Pilot-Scale Anaerobic Co-Digestion of Sewage Sludge with Agro-Industrial by-Products for Increased Biogas Production of Existing Digesters at Wastewater Treatment Plants." *Waste Management* 59: 362–370. doi:10.1016/j.wasman.2016.10.043.

Misi, S. N., and C. F. Forster. 2001. "Batch Co-Digestion of Multi-Component Agro-Wastes." *Bioresource Technology* 80 (1). Elsevier: 19–28. doi:10.1016/S0960-8524(01)00078-5.

Nuchdang, Sasikarn, and Chantaraporn Phalakornkule. 2012. "Anaerobic Digestion of Glycerol and Co-Digestion of Glycerol and Pig Manure." *Journal of Environmental Management* 101: 164–172. doi:10.1016/j.jenvman.2012.01.031.

Perry, Rh, Dw Green, and Jo Maloney. 1997. *Perry's Chemical Engineers' Handbook. Journal of Chemical Education.* doi:10.10360071422943.

Ripley, L E, W C Boyle, and J C Converse. 1986. "Improved Alkalimetric Monitoring for Anaerobic Digestion of High-Strength Wastes." *Water* 58: 406–11. doi:10.1016/S0262-1762(99)80122-9.

Selma, Vivian C., Luís H. B. Cotrim, José A. D. Rodrigues, Suzana M. Ratusznei, Marcelo Zaiat, and Eugenio Foresti. 2010. "ASBR Applied to the Treatment of Biodiesel Production Effluent: Effect of Organic Load and Fill Time on Performance and Methane Production." *Applied Biochemistry and Biotechnology* 162 (8): 2365–2380. doi:10.1007/s12010-010-9009-x.

Siles, J. A., M. A. Martín, A. F. Chica, and A. Martín. 2010. "Anaerobic Co-Digestion of Glycerol and Wastewater Derived from Biodiesel Manufacturing." *Bioresource Technology* 101 (16): 6315–6321. doi:10.1016/j.biortech.2010.03.042.

Silva, Renato C., José A D Rodrigues, Suzana M. Ratusznei, and Marcelo Zaiat. 2013. "Anaerobic Treatment of Industrial Biodiesel Wastewater by an ASBR for Methane Production." *Applied Biochemistry and Biotechnology* 170 (1): 105–118. doi:10.1007/s12010-013-0171-9.

Vlassis, T., K. Stamatelatou, G. Antonopoulou, and G. Lyberatos. 2013. "Methane Production via Anaerobic Digestion of Glycerol: A Comparison of Conventional (CSTR) and High-Rate (PABR) Digesters." *Journal of Chemical Technology and Biotechnology* 88 (11): 2000–2006. doi:10.1002/jctb.4059.

Zaiat, M., A. K. A. Cabral, and E Foresti. 1994. "Horizontal-Flow Anaerobic Immobilized Sludge Reactor for Wastewater Treatment: Conception and Performance Evaluation." *Brazilian Journal of Chemical Engineering* 11: 33–42.

INDEX

D

E

F

G

N

O

P

Q

R

S

T

U

V

W

Y